Membrane Reactors for Hydrogen Production Processes

Marcello De Falco · Luigi Marrelli ·
Gaetano Iaquaniello
Editors

Membrane Reactors for Hydrogen Production Processes

Editors
Dr. Marcello De Falco
Faculty of Engineering
University Campus Bio-Medico of Rome
via Alvaro del Portillo 21
00128 Rome
Italy
e-mail: m.defalco@unicampus.it

Dr. Gaetano Iaquaniello
Tecnimont-KT S.p.A.
Viale Castello della Magliana 75
00148 Rome
Italy
e-mail: Iaquaniello.G@tecnimontkt.it

Prof. Luigi Marrelli
Faculty of Engineering
University Campus Bio-Medico of Rome
via Alvaro del Portillo 21
00128 Rome
Italy
e-mail: l.marrelli@unicampus.it

ISBN 978-1-4471-6050-2 ISBN 978-0-85729-151-6 (eBook)

DOI 10.1007/978-0-85729-151-6

Springer London Dordrecht Heidelberg New York

British Library Cataloguing in Publication Data
A catalogue record for this book is available from the British Library

Cover design: eStudio Calamar S.L.

Printed on acid-free paper

Springer is part of Springer Science+Business Media (www.springer.com)

Foreword

A significant increase of interest in membrane reactors in recent years together with several research programs which are in progress nowadays has pushed the authors to consolidate their experiences into this book focusing on the more attracting and promising processes. Few authors have been working together in an R&D project started in 2006. The goal of this project was the development of a new scheme for hydrogen production through stream reforming based on integrating chemical reaction and membrane separation.

One of the main objectives of the project was to build a pilot plant with industrial-size components. Today the pilot plant (see figure below) is running and the experimental campaigns are allowing to verify the reliability of the novel scheme and at the same time useful information is generated to assess the plant operation and for scaling-up the design to future industrial size units.Semi-industrial membrane-based steam reformer installed in Chieti Scalo (Italy)

The R&D project was financially supported by the Italian Research Ministry, and managed by the Chemical Engineering Department of the University of L'Aquila.

In the next year the experimental program will be oriented both to the theoretical aspects of the process and to the technological ones, focusing on the reliability and on the scaling-up criteria for process design of industrial plants.

The research project allowed to compare two different ways of integrating catalytic reaction and hydrogen separation: one in which the membranes are set inside the reaction environment (e.g. the membranes constitute the catalytic tube walls) and another one in which the catalytic module is separated from the hydrogen permeation one.

With reference to the book content, the authors divided it into 11 chapters. The book starts with an overview on membrane selective membranes integrated in the chemical reaction environment. The thermodynamics and kinetics of membrane reactors are also formulated and assessed for different membrane reactor architectures.

The rest of the book is divided into three parts. The first part deals with the membranes, membranes manufacturing and mathematical modeling. The second

Semi-industrial membrane-based steam reformer installed in Chieti Scalo (Italy)

reviews the most attracting application from an industrial point of view. The third is dedicated to the description of the pilot plant where the novel configuration was implemented at a semi-industrial scale.

In Chap. 2, an extensive overview concerning palladium-based membranes is presented. In particular, an assessment of the problems associated with palladium membranes is given followed by a description of the preparation methods. Chapter 3 focuses on a proper membrane manufacturing strategy to improve their industrial competitiveness by lowering production costs. In Chap. 4, the mathematical modeling strategies focused on the simulations of membrane reactors (MR) and reactor membrane modules (RMM) are presented. In Chap. 5, the membrane reactor application to natural gas steam reforming is presented and assessed in the MR and in the RMM configuration. Chapter 6 is dedicated to the autothermal reforming reaction (ATR) for syngas production. An higher efficiency can be achieved integrating in the reactor an H_2 permselective membrane. In Chap. 7, the water–gas shift reaction is studied. The integration of membranes inside or outside the shift environment is also analyzed. A techno-economic analysis is dedicated to the shift reactor integrated with the membrane module compared to the conventional process. In Chap. 8 the H_2S catalytic cracking process is analyzed. Also in this case the benefits derived by coupling the Claus reaction and the hydrogen module separation are examined with reference to different process configurations. In particular, the high Claus reaction temperature (900°C) and the low working temperature of ceramic membrane modules (600°C) are taken into account in the definition of process multi-staged configuration. Chapter 9 deals with alkanes dehydrogenation. Olefins production by catalytic dehydrogenation of light alkanes might be an alternative to conventional heavy hydrocarbons cracking. Also in this case the catalytic membrane reactor might improve the process yield. In particular is studied the coupling of the Pd–Ag membrane and catalysts usually used in the dehydrogenation process that need to operate at low H_2 partial pressures. In Chap. 10 the characteristics of the pilot plant discussed at the beginning of this foreword are shown in more detail referred to the process design and to the first positive experimental results.

Each chapter was developed as a whole that can be read without reference to the others.

Professor Diego Barba
Scientific Director

Contents

List of Contributors

A. Basile, Institute on Membrane Technology of National Research Council (ITM-CNR), Via P. Bucci Cubo 17/C c/o University of Calabria, Rende (CS), 87036, Italy

A. Borruto, Department of Chemical Engineering, Materials and Environment, University of Rome La Sapienza, via Eudossiana 18, 00184, Rome, Italy

E. Brancaccio, Processi Innovativi S.r.l., Corso Federico II 36, 67100, L'Aquila, Italy

P. Ciambelli, Department of Industrial Engineering, University of Salerno, 84084, Fisciano (SA), Italy

Marcello De Falco, Faculty of Engineering, University Campus Bio-Medico of Rome, via Alvaro del Portillo 21, 00128, Rome, Italy

J. Galuszka, Natural Resources Canada, CanmetENERGY, 1 Haanel Drive, Ottawa, ON, K1A 1M1, Canada

T. Giddings, Natural Resources Canada, CanmetENERGY, 1 Haanel Drive, Ottawa, ON, K1A 1M1, Canada

G. Iaquaniello, Tecnimont KT, Viale Castello della Magliana 75, 00148, Rome, Italy

A. Iulianelli, Institute on Membrane Technology of National Research Council (ITM-CNR), Via P. Bucci Cubo 17/C c/o University of Calabria, Rende (CS), 87036, Italy

D. Katsir, Acktar Ltd., 1 Leshem St, P.O.B. 8643, Kiryat-Gat, 82000, Israel

S. Liguori, Institute on Membrane Technology of National Research Council (ITM-CNR), Via P. Bucci Cubo 17/C c/o University of Calabria, Rende (CS), 87036, Italy

T. Longo, Institute on Membrane Technology of National Research Council (ITM-CNR), Via P. Bucci Cubo 17/C c/o University of Calabria, Rende (CS), 87036, Italy

E. Lollobattista, Processi Innovativi S.r.l., Corso Federico II 36, 67100, L'Aquila, Italy

L. Marrelli, Faculty of Engineering, University Campus Bio-Medico of Rome, via Alvaro del Portillo 21, 00128, Rome, Italy

G. Narducci, Department of Chemical Engineering, Materials and Environment, University of Rome La Sapienza, via Eudossiana 18, 00184, Rome, Italy

V. Palma, Department of Industrial Engineering, University of Salerno, 84084, Fisciano (SA), Italy

E. Palo, Tecnimont KT, Viale Castello della Magliana 75, 00148, Rome, Italy

A. Salladini, Processi Innovativi S.r.l., Corso Federico II 36, 67100, L'Aquila, Italy

M. Sheintuch, Technion, Technion City, 32000, Haifa, Israel

D. S. A. Simakov, Technion, Technion City, 32000, Haifa, Israel

Chapter 1
Integration of Selective Membranes in Chemical Processes: Benefits and Examples

Luigi Marrelli, Marcello De Falco and Gaetano Iaquaniello

1.1 Introduction

Integration of reaction and separation in a single unit is a powerful tool to increase efficiency and economic advantages of many chemical processes. Reactive distillation, extraction, and adsorption are well-known examples of this technological resource. Recently, a very promising solution is offered by membrane reactors (MRs).

A MR is a system coupling reaction and separation of one or more products, with the separation operation performed by a selective membrane. Although not yet very used at industrial scale, MRs are attracting the attention of scientists and engineers in the last two decades, and many interesting articles have appeared in the literature on their performance and possible application in many fields of chemical and biochemical industries. An excellent review about these topics has appeared in 2002 by Sanchez Marcano and Tsotsis [1].

For biotechnological applications, synthetic membranes entrapping enzymes, bacteria, or animal cells are used in membrane bioreactors disclosing new important developments mainly due to the increased stability of immobilized enzymes, the possibility of their continuous reuse and the absence of pollution of the products. Membrane bioreactors are of great interest as well for the possibility of continuously removing metabolites whose presence in the reaction environment could reduce the productivity of the reactor.

In the field of inorganic heterogeneous catalysis, metallic or ceramic membranes are used to bear the generally more severe thermal conditions. Both dense and

L. Marrelli (✉) and M. De Falco
Faculty of Engineering, University Campus Bio-Medico of Rome,
via Alvaro del Portillo 21, 00128 Rome, Italy
e-mail: l.marrelli@unicampus.it

G. Iaquaniello
Tecnimont KT, Viale Castello della Magliana 75, 00148 Rome, Italy

M. De Falco et al. (eds.), *Membrane Reactors for Hydrogen Production Processes*,
DOI: 10.1007/978-0-85729-151-6_1,

porous, inert and catalytically active membranes have been used in the different processes analyzed in the scientific or technical literature. Dense Pd-based membranes or almost dense SiO_2 membranes offer very good selectivity and permeability features in all reactions involving generation or consumption of hydrogen [2]. Composite membranes made of a dense selective layer supported on a porous material can represent a technical solution when strong mechanical stresses are imposed to the membrane. Palladium and its alloy membranes were prepared on stainless steel supports by the electroless plating technique and on alumina supports by sputtering or by metal–organic chemical vapor deposition [3–6].

Two main advantages are offered by MRs. If the membrane is very selective with regard to a specific product, then it is possible to obtain a very pure compound in the same equipment used to produce it. Furthermore, in the case of reversible reactions, removing one or more reaction products as they are generated allows conversions higher than equilibrium values to be reached.

The use of inorganic MRs has been investigated for a number of reactions.

Pd-based membranes, mainly Pd–Ag (23%wt), have been extensively tested for natural gas steam reforming, which is the main process to produce large amount of hydrogen. Many experimental works are reported in the literature [7–11], attesting the good performance in terms of natural gas conversion at much lower operating temperature than traditional process (methane conversions up to 90–95% at 450–550 vs. 850–1000°C).

The dehydrogenation of cyclohexane to benzene [12] and of ethylbenzene to styrene [13, 14] have been studied in MRs using glass or alumina membranes. Even if a conversion increase beyond the equilibrium value has been observed in all the cases, compared with Pd-based membranes, which are much more selective with respect to hydrogen, porous membranes are less efficient in improving the conversion. For example, the conversion of cyclohexane to benzene at 200°C and 1 atm (equilibrium conversion 19%) is 45% in a Vycor glass MR whereas it becomes 99.7% in a Pd–Ag MR.

A growing interest in the membrane assisted Water–Gas Shift reaction (WGSR) is manifested by a substantial volume of the published pertinent literature that could be grouped around the two most popular hydrogen permselective membranes, based either on palladium [15–19] or silica [20–22]. Among them, Basile et al. [15] studied the WGSR in a palladium MR and showed the importance of the membrane preparation method in obtaining high-quality membrane materials. The best membrane increased CO conversion up to 99.89% at about 330°C with a performance claimed stable for more than 2 months [16].

The most recent application of a silica membrane—prepared by counter-diffusion CVD of TMOS (TetraMethylOrthoSilicate) and oxygen—to H_2S decomposition was reported by Akamatsu et al. [23]. It was claimed that using this membrane and a commercial desulphurization catalyst at 600°C, about 70% H_2S diluted in 99% nitrogen was converted in a relatively short residence time of 7 s. A similar application but with a different process architecture was proposed by Galuzka et al. [24] and is reported in Chap. 8.

In general, membrane integration inside the reaction environment involves the following main benefits:

1. The ability of such a reactor to circumvent thermodynamic limitation of an equilibrium-controlled process allows the same reactants conversion to be obtained at a lower temperature or a higher conversion to be reached at the same temperature.
2. Operating at lower reaction temperatures, new heat integration strategies to supply heat duty to reactors can be proposed. The use of gas exhausts from a gas turbine as suggested by [25] or solar heated molten salts [26] could become applicable and reliable industrial solutions.
3. Lower operating temperatures reduce materials cost as well as increase operation safety.
4. The expected significant process simplification and intensification would capitalize on a new industrial paradigm offered by equipments combining reaction and separation in one step. New reactors and overall plant design strategy have to be defined.

In this chapter, a theoretical demonstration of operating benefits for chemical processes in integrating selective membrane is given to the readers. For practical cases refer to the following chapters.

1.2 Thermodynamics and Kinetics Background

Reversible reactions are thermodynamically limited since equilibrium conditions cannot be overcome in the reacting mixture. From a thermodynamic point of view, equilibrium is represented by a constraint (equilibrium constant) on mole fractions (or concentrations), temperature, and pressure; this constrain derives from the second principle of thermodynamics. At equilibrium conditions, no net change in state variables is observed.

From a kinetic point of view this means that at equilibrium the reaction rate of the direct reaction is equal to the reaction rate of the inverse reaction. As known, the reaction rates of the direct and of the inverse reactions generally increase with the reactant and the product concentrations, respectively. Only when product concentrations are high enough and reactant concentrations are low enough, i.e., when the conversion of a key reactant is high enough, the equilibrium is reached. Some kinds of reactions (irreversible reactions) require very high conversions to reach equilibrium conditions at a given temperature, whereas for other reactions a quite low value of the conversion corresponds to equilibrium. These latter reactions are named reversible reactions according to the thermodynamic statement that each transformation close to equilibrium is a reversible transformation. In the case of reversible reactions, the thermodynamic limit can prevent acceptable conversions to be obtained.

The equilibrium composition depends on temperature and pressure of the system as well as on the composition of the charge or feed.

A very simple numerical example can be useful to illustrate these subjects.

Let's consider the following synthesis reaction in gas phase:

$$A + B \Leftrightarrow R$$

Let be $K = 1.5$ the equilibrium constant, at a given temperature T. At a not too high pressure P, the equilibrium composition in terms of mole fraction is easily obtained from the definition of K:

$$K = \frac{y_R}{y_A \cdot y_B} \cdot P^{-1} \tag{1.1}$$

If the charge is composed of $n_A^0 = 5$ mol of A and $n_B^0 = 5$ mol of B, equilibrium conversion of A at $P = 1$ atm, is $X_A = 0.367$.

This conversion depends on the charge composition. For example, with $n_A^0 = 5$ and $n_B^0 = 3$, at the same temperature and pressure, we get $X_A = 0.274$ at equilibrium, whereas with $n_A^0 = 3$ and $n_B^0 = 5$ we get $X_A = 0.399$.

The effect of the pressure depends on the phase of reacting system and on the stoichiometry of the reaction. In the present case of reaction in gas phase with a negative sum of stoichiometric coefficients, high values of P promote the conversion.

In any case, higher values of equilibrium conversion can be obtained with a higher value of the equilibrium constant. For example, with $K = 1.5$, $P = 15$ atm and $n_A^0 = n_B^0 = 5$ we get $X_A = 0.75$.

Since the equilibrium constant K depends on the temperature, a way to change its value is to change T. In particular, K increases with T for endothermic reactions and decreases for exothermic reactions according the well-known Van't Hoff equation:

$$\frac{d \ln K}{dT} = \frac{\Delta H}{R \cdot T^2} \tag{1.2}$$

with $\Delta H > 0$ for endothermic reactions and $\Delta H < 0$ for exothermic reactions.

In the case of exothermic reactions, low temperatures should be required to get acceptable equilibrium conversions but the corresponding slow reaction rates should require very high residence time in the reactor.

On the contrary, in the case of endothermic reactions, suitable values of K are often obtained only at very high temperatures with corresponding unacceptable heat duties, quick deactivation of catalysts, and technical and economic problems in construction materials.

These shortcomings can be overcome by removing one or more products from the reaction environment to prevent the equilibrium composition can be reached, i.e. maintaining y_R in Eq. 1.1. at lower values than equilibrium.

1.2.1 Thermodynamics of Reacting Systems

It is well known that a way to express equilibrium condition in a closed system is to set Gibbs free energy at its minimum value at constant P and T.

$$\mathrm{d}G = 0 \quad [P, T] \tag{1.3}$$

Equation 1.3, that is a consequence of the second Principle of Thermodynamics, states that any change from the equilibrium state at constant T and P involves an increase of G.

If the system is a multi-component reacting system, Eq. 1.3 can be expressed in the following form:

$$\sum_{i=1}^{c} \alpha_i \mu_i = 0 \tag{1.4}$$

where α_i are the stoichiometric coefficients of the components (assumed to be positive for products and negative for reactants) and μ_i are the chemical potentials defined as:

$$\left(\frac{\partial G}{\partial n_i}\right)_{P,T,nj \neq i} = \mu_i \tag{1.5}$$

If chemical potentials at system conditions are expressed in terms of fugacities f_i, and components are assumed to be pure in the reference state, Eq. 1.4 becomes:

$$-\frac{\Delta g^0}{R \cdot T} = \ln \prod \left(\frac{\bar{f}_i(P, T, z_i^{\mathrm{eq}})}{f_i^0(P^{\oplus}, T)}\right)^{\alpha i} \tag{1.6}$$

where $\Delta g^0 = \sum \alpha_i \cdot g_i^0$, g_i^0 are the molar free energies of pure components and z_i^{eq} are the mole fractions at equilibrium conditions.

The product at the r.h.s. of Eq. 1.6 is the thermodynamic constant K of the reaction.

$$K(P^{\oplus}, T) = \prod \left(\frac{\bar{f}_i(P, T, z_i^{\mathrm{eq}})}{f_i^0(P^{\oplus}, T)}\right)^{\alpha i} \tag{1.7}$$

Therefore, Eq. 1.6 can be written in the following form:

$$\ln K(P^{\oplus}, T) = -\frac{\Delta g^0}{R \cdot T} \tag{1.8}$$

The value of K depends on the temperature and on the pressure and composition conditions assumed as reference state. Once pressure and composition are fixed, equilibrium constant depends on the temperature only.

1.2.2 Affinity and Evolution of Reacting Systems

If a closed system is not at equilibrium, a change of its state variables is observed. The direction of the system evolution is stated by the second Principle of Thermodynamics and corresponds to a positive production of entropy $d_iS > 0$:

$$d_iS = -\sum_{j=1}^{c} \frac{\mu_j}{T} dn_j \geq 0 \tag{1.9}$$

where the equality holds at equilibrium conditions. The presence of chemical reactions in a closed system causes changes in the mole number of reactants and products which are not independent each other since the stoichiometry of reactions imposes some constraints.

In the case of r simultaneous reactions and c components, we have:

$$dn_j = \sum_{k=1}^{r} \alpha_{j,k} \cdot d\xi_k \qquad j = 1 \ldots c \tag{1.10}$$

where ξ_k is the degree of advancement of reaction k and $\alpha_{j,k}$ indicates the stoichiometric coefficient of component j in reaction k.

Using conditions 1.10 in Eq. 1.9 we get:

$$d_iS = \frac{1}{T} \sum_{k=1}^{nr} A_k d\xi_k \geq 0 \tag{1.11}$$

or in terms of entropy production rate:

$$\frac{d_iS}{dt} = \frac{1}{T} \sum_{k=1}^{nr} A_k \frac{d\xi_k}{dt} \geq 0 \tag{1.12}$$

In Eqs. 1.11 and 1.12, A_k is the affinity of reaction k defined as follows:

$$A_k = -\sum_{j=1}^{c} \alpha_{jk} \cdot \mu_j \tag{1.13}$$

In the case of a single reaction, Eq. 1.12 states that the reaction proceeds in the direction of increasing ξ when the affinity is positive while ξ decreases for $A < 0$. At equilibrium, the affinity of the reaction is zero:

$$A = -\sum_{j=1}^{c} \alpha_j \mu_j = 0 \qquad \text{Equilibrium condition} \tag{1.14}$$

In the case of multiple reactions, Eq. 1.12 is satisfied even with some terms A_k $d\xi_k < 0$ if the summation is greater than 0. Obviously, at equilibrium conditions, since $d_i S$ must be zero for every infinitesimal variation of degrees of advancement, the affinity of each reaction has to be zero:

$$A_1 = A_2 = \cdots A_{nr} = 0 \tag{1.15}$$

At non-equilibrium conditions, the reaction proceeds from left to right ($d\xi > 0$) only if $A > 0$.

This means (Eqs. 1.6 and 1.13):

$$\begin{aligned} &\left[\Delta\mu^{\oplus} + R \cdot T \cdot \ln \hat{K}\right] < 0 \\ &R \cdot T \cdot \left[\ln \hat{K} - \ln K\right] < 0 \end{aligned} \tag{1.16}$$

where:

$$\hat{K} = \prod \left(\frac{\bar{f}_i(P, T, z_i)}{f_i^{\oplus}(P^{\oplus}, T)} \right)^{\alpha i} \tag{1.17}$$

In other terms, if $\hat{K} < K$ mole fractions of products are lower than equilibrium values and mole fractions of reactants are higher than equilibrium values and the reaction evolves from left to right.

Vice versa, if $\hat{K} > K$ the evolution is from right to left. Finally, if $\hat{K} = K$, the reaction is at equilibrium and its evolution stops.

In each case, outside equilibrium condition, the reaction evolves in the direction corresponding to a decrease of free energy G of the system.

Selective membrane application allows products mole fraction to be reduced and consequently the condition $\hat{K} < K$, i.e., mole fractions of products lower than equilibrium values, is verified, promoting continuously the reaction from left to right.

1.2.3 Kinetics of Reversible Reactions

The reaction rate of a reversible reaction is the result of two contributions: the forward reaction that generates the products (chemical species at the right of the stoichiometric equation) from the reactants (chemical species at the left of the stoichiometric equation) and the reverse reaction that proceeds in the opposite direction.

For instance, in the case of a general reaction:

$$\alpha_1 A_1 + \alpha_2 A_2 \Leftrightarrow \alpha_3 A_3 + \alpha_4 A_4 \tag{1.18}$$

the net disappearance rate of A_1 is given by:

$$-r_1 = r_{1,\text{forward}} - r_{1,\text{reverse}} \tag{1.19}$$

where the forward reaction rate generally increases with the concentration of reactants A_1 and A_2 whereas the reaction rate of the reverse reaction generally increases with the concentration of the products A_3 and A_4.

If the concentrations of the reactants and of the products are such as the forward reaction rate is higher than the reverse reaction rate, the reaction evolves from left to right; vice versa the direction of the reaction is from "products" to "reactants" when the product and reactant concentrations make the reverse reaction faster than the forward reaction.

At equilibrium, no net change is observed in the reaction mixture so that forward and reverse reaction rates are equal.

Removing a product from the reaction environment usually reduces the reaction rate of the reverse reaction allowing the reaction to evolve further from left to right.

1.3 Membrane Reactors

Thermodynamics and kinetics conclusions reported in the previous sections show that removing one or more products from the reaction environment allows the reaction to proceed without reaching equilibrium conditions.

A technical way to fulfill this plan is to surround the reacting mixture by a membrane selectively permeable to one of the products at least (Integrated Membrane Reactor). As a result, the reaction rate of the reverse reaction is lower than that of the forward reaction and the conversion increases theoretically to the value corresponding to the complete depletion of the limiting reactant.

A tubular MR is simply composed of two co-axial tubes with the inner one made of a material selectively permeable to a product of the reaction.

The reacting stream can be fed to the annular region or to the inner tube. The region where the reactants are fed and the reaction takes place is called *reaction* or *retentate* region whereas the part of the reactor where the products permeated through the membrane are collected is the *permeate* chamber.

In the permeate region, the products are usually swept by a gas easily separable from them (often water vapor). The sweep gas can move co-currently or counter-currently with respect to the reacting mixture. Each of these schemes has advantages and drawbacks.

A schematic representation of these two possibilities is shown in Fig. 1.1. The reaction chamber is the inner tube where a fixed bed of catalyst is assumed to be present. The permeate chamber is the annular region between the inner tube and the shell. In the permeate region, the sweep gas flows in co-current mode in the case (a) and in counter-current mode in the case (b). One or more products of the reaction are shown to flow through the membrane from the catalytic-fixed bed toward the permeate region. However, in co-current scheme an inversion of the flux is theoretically possible at some distance from the inlet section since the flux

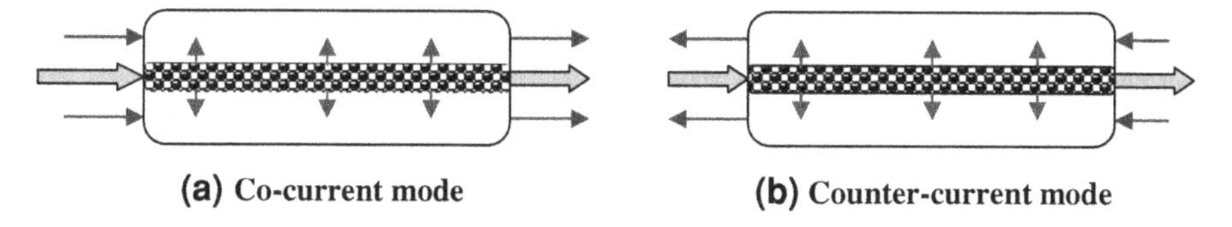

Fig. 1.1 Schemes of feed and sweep gas motion

of a compound occurs from regions with high partial pressures toward those with lower partial pressures. In co-current mode, both reacting mixture and sweep gas are relatively poor of products in the inlet section so that some compounds generated by the reaction can diffuse from the reactor to the permeate chamber; but, along the axis of the reactor the sweep gas is more and more rich of these compounds where their partial pressure in the reacting mixture decreases due to the transport through the membrane, the slowing of the reaction rate and the pressure drop along the catalytic bed. When partial pressures have the same value in the reaction and permeate chambers, the flux stops. In the next sections of the reactor, an inverse transport from permeate to the reaction region could occur if the effect of pressure drop is prevailing on the generating rate due to the reaction.

This anomalous behavior is to be taken into account, but it is usually only a theoretical oddness.

Obviously, this anomaly is not generally allowed in counter-current mode where the end section of the reactor is characterized by low pressures of removable compounds in the reacting region but by a practically pure sweep gas entering into the permeate region.

In Fig. 1.2, a draft of a reactor for methane steam reforming with multiple membrane tubes for hydrogen separation, operating in co-current mode, is shown.

Unfortunately, the integrated solution of MRs, although very intriguing for the compactness of the equipments, presents some technical problems.

Some membranes are sensitive to high temperatures so that they could be incompatible with the reactor temperature. Furthermore, in an integrated MR, damage to the membrane needs to stop the system, to unload the catalyst (if present), to substitute the membrane, and a new start-up of the system.

Due to these main reasons, new configurations called Staged Membrane Reactor (SMR) or Reformer and Membrane Modules (RMMs) have been proposed in the technical literature [25–29].

A SMR is a series of modules (RMM) each composed of a traditional reactor followed by a membrane separation unit. The stream flowing out of the reactor, rich of reaction products, enters into the separation unit where the selective membrane removes one of the products. The retentate of the membrane unit is then fed to the subsequent module. A system of heat exchangers can be present between each unit and the following one. A system composed of two modules is shown in Fig. 1.3.

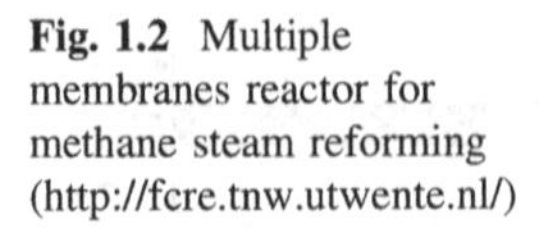

Fig. 1.2 Multiple membranes reactor for methane steam reforming (http://fcre.tnw.utwente.nl/)

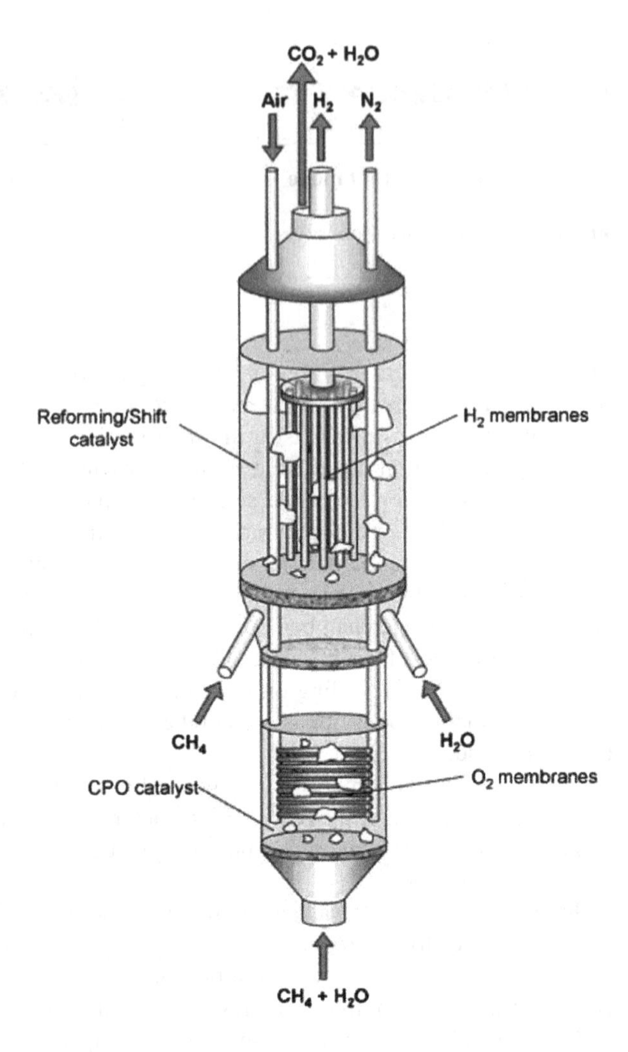

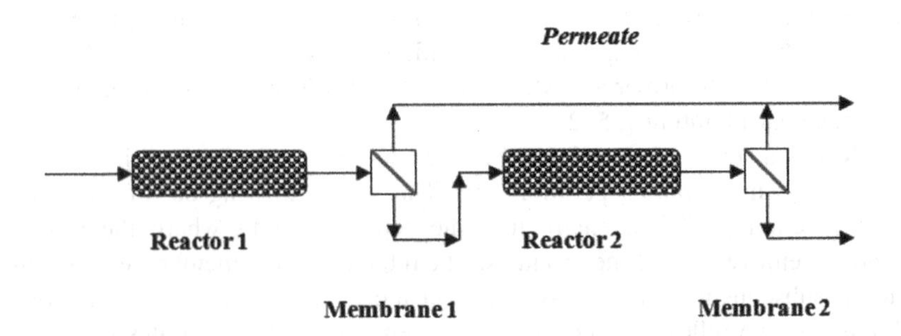

Fig. 1.3 Staged membrane reactor with two modules

A SMR is not strictly a single equipment but each module and the whole system offer "enhanced performance in terms of separation, selectivity, and yield" [1].

An optimized design of a SMR requires the evaluation of a number of parameters such as number of modules, conversion to be achieved in each reactor, degree of product removal in each membrane unit, temperature distribution etc.

It is easy to realize that a system composed of an infinite number of modules approaches the behavior of an integrated MR if the reaction volume and the product removal degree are infinitesimal in each module.

Figure 1.4 shows the RMM test plant having the capacity of 20 Nm^3/h of pure hydrogen, developed and fabricated by Tecnimont KT and deeply described in Chap. 10. The plant, composed by a two-stages natural gas steam reformer + Pd-based separation modules, has demonstrated the feasibility of SMR configuration.

Fig. 1.4 20 Nm^3/h of pure H2 capacity RMM plant (Courtesy from Tecnimont KT)

1.4 Theoretical Study of Staged Membrane Reactor Performance

The following very simple reversible reaction in gaseous phase can be used to show the effect of separation units on the attainable conversion:

$$\mathrm{A} \Leftrightarrow \mathrm{B} \qquad (-r_\mathrm{A}) = k_1 \cdot P_\mathrm{A} - k_2 \cdot P_\mathrm{B} \tag{1.20}$$

If the equilibrium constant K_P at a certain temperature T is assumed to be $K_P = 0.4$, the equilibrium conversion is:

$$X_\mathrm{A}^{\mathrm{eq}} = \frac{K_P}{K_P + 1} = 0.8$$

This is the maximum value of conversion attainable in an isothermal reactor operating at the temperature T. It is easy to show that the reaction volume needed to achieve a conversion X_A in an isobaric and isothermal plug flow reactor is:

$$\mathrm{V}(X_\mathrm{A}) = \frac{F_\mathrm{A}^0}{k_1 \cdot P} \cdot X_\mathrm{A} \cdot \ln\left(\frac{X_\mathrm{A}^{\mathrm{eq}}}{X_\mathrm{A}^{\mathrm{eq}} - X_\mathrm{A}}\right) \tag{1.21}$$

In order to achieve the equilibrium conversion an infinite volume V of reactor is required since forward and reverse reaction rates tend to become equal as equilibrium is approached.

However, if the scheme of Fig. 1.3 is adopted and a membrane unit (Membrane 1) is used to remove a fraction γ of the product B generated in the first reactor of finite volume V_1, the reaction can proceed in the second reactor and a conversion even higher than the equilibrium value can be reached in a finite reaction volume given by the sum of the two reactor volumes, V_1 and V_2.

In Fig. 1.5, the reaction volume required to achieve a conversion X_A is shown for three values of removal fraction γ.

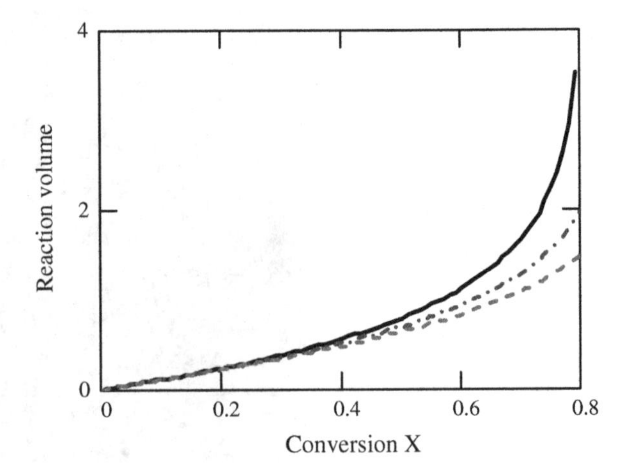

Fig. 1.5 Reaction volume versus conversion: $\gamma = 0$ *solid line*; $\gamma = 0.3$ *dash and dot line*; $\gamma = 0.5$ *dash line* (parameters used in this simulation: $k_1 = 1$ mol/(m^3 min atm); $P_1 = P_2 = 5$ atm; $F_\mathrm{A}^0 = 5$ mol/min.)

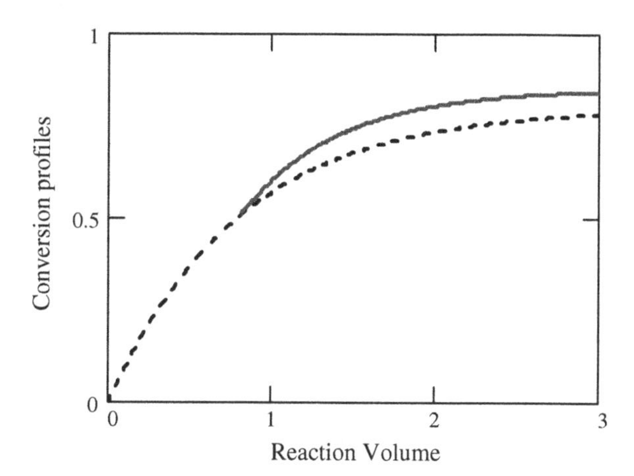

Fig. 1.6 Conversion profiles along the reactors. *Dash line*: only a reactor without product removal; *solid line*: 50% product removal at $V_1 = 0.8$

An infinite volume is required to get $X_A = 0.8$ if no product removal is implemented, whereas the same conversion can be obtained by very lower volumes (about $V = 2$ and $V = 1.5$) when removal ratios ($\gamma = 0.3$ and 0.5, respectively) are used.

Likewise, in a reaction volume of 1, a conversion $X_A = 0.572$ is achieved when no product removal is carried out, while conversions $X_A = 0.625$ and $X_A = 0.674$ are obtained with removal fractions $\gamma = 0.3$ and $\gamma = 0.5$, respectively.

Figure 1.6 shows the effect of product removal on conversion profiles along the first and the second reactor (see Fig. 1.3).

A removal ratio $\gamma = 0.5$ used in the first membrane unit (Fig. 1.3) located at the outlet of a reactor with a volume $V_1 = 0.8$ allows the conversion profile in the second reactor to be higher than in the case of absence of membrane unit.

1.5 Recycle of Retentate

The fluid not permeated at the end of the set of modules contains reaction products as well as un-reacted compounds which could be a valuable material. This mixture undergoes different treatments depending on the nature of the material and the kind of the process. In some cases, a further separation and purification operation are used after the reaction step to recover useful compounds. For example, in the case of steam reforming of light hydrocarbons for producing hydrogen, a pressure swing adsorption (PSA) step could be used for recovering pure hydrogen not permeated through the membrane of the reaction system.

An interesting solution is used in cogeneration processes when the retentate gas has a high energetic content. In this case, a part of the exiting gas is burned to give power to an electric generator whereas the residual part is recycled to the reaction system to convert the un-reacted compounds. A simple scheme of the recycle system is shown in the Fig. 1.7, while a complete process scheme of a two steps

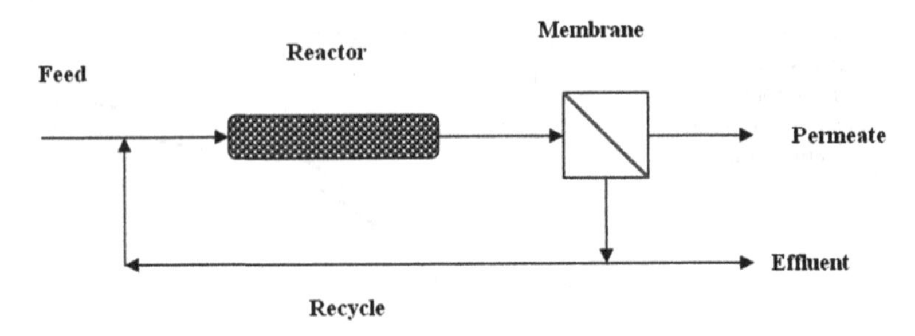

Fig. 1.7 Membrane system with recycle

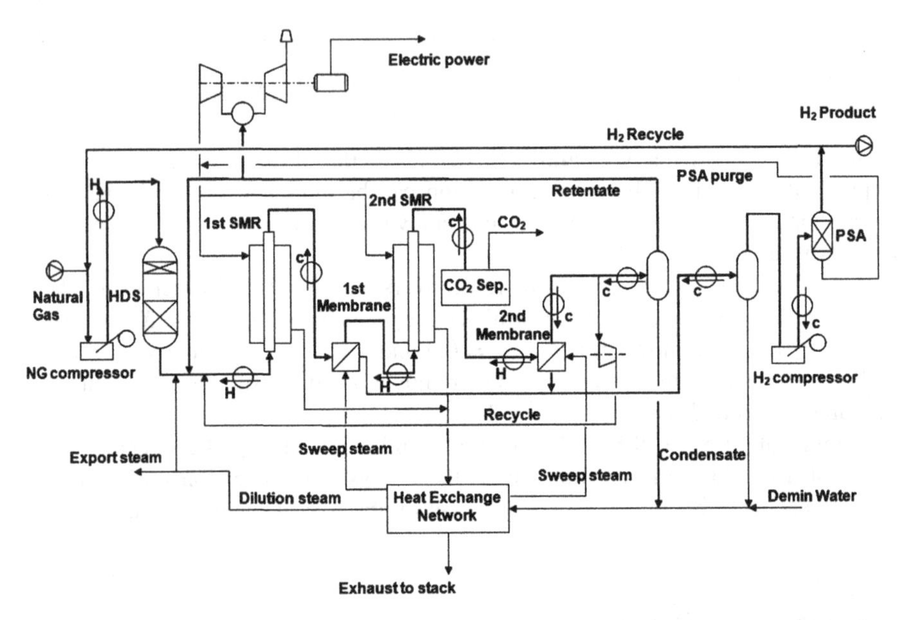

Fig. 1.8 Two-steps RMM natural gas steam reforming process with electric power production and recycle layout [25, 29]

RMM plant with a recycle stream and electricity power production by retentate gas is reported in Fig. 1.8 (refer to [25, 29] for a complete description and assessment of the plant).

It is known that, in a traditional tubular reactor, the recycle of a portion of the exiting stream makes worse the performance of the reactor; however, the case of MRs is very different due to the removal of a product and considerable improvements can be achieved.

Figure 1.9 shows the effect of recycle ratio R on the conversion.

The conversion X_A is defined as the reacted fraction of A in the feed leaving the system with the effluent. The removal percentage ρ is the ratio between the flow rate of B in permeate and the flow rate of this product entering the membrane module.

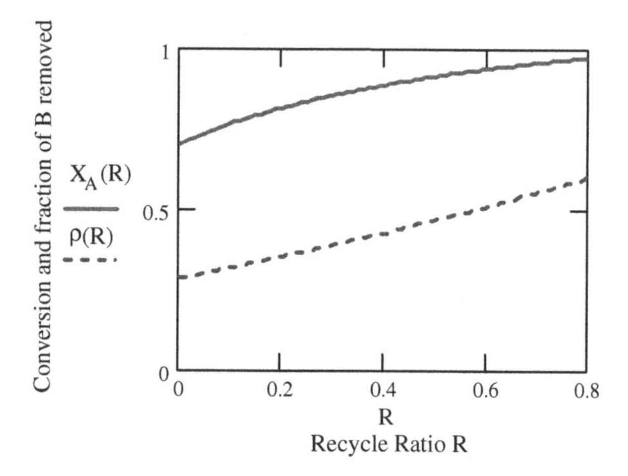

Fig. 1.9 Improving effect of recycle ratio R on the conversion and fraction of B to be removed in the membrane module at a fixed flow rate of B in the permeate stream

Table 1.1 Reference parameters used in calculations

Reactor volume V, m^3	Flow rate of A in the feed F_A^0, mol/min	Pressure P, atm	Kinetic constant of forward reaction k_1, $\frac{mol}{m^3\ min\ atm}$
1.2	5	5	1
Kinetic constant of forward reaction k_2, $\frac{mol}{m^3\ min\ atm}$			Flow rate of B in the permeate $F_{B,perm}$, mol/min
2.5			1

Calculations have been performed at the indicative values of parameters reported in Table 1.1 with the simplifying assumptions of isothermal and isobaric reactor and infinite selectivity of membrane.

Fixing the flow rate of B in the permeate is a realistic basis since in many cases the aim in using a membrane system with recycle is just to obtain a specific flow rate of the product with a high conversion of the reactant. Obviously, the dependence on the recycle ratio R of the fraction of B to be removed in the membrane requires a suitable design of the membrane module which depends on the adopted value of R.

1.6 Design of Membrane Module

The flux J_B through a membrane layer depends on the properties of the layer material and of the compound B, on the thickness of the membrane and on the difference of partial pressures between the two chambers of the module. The properties of materials are taken into account by means of the membrane permeability which is affected by the temperature as well. A detailed description of mass transport through Pd-based membrane is given in Chap. 2 of this book. When Sieverts-Fick law is valid, the flux of B is given by:

$$J_B = \frac{Pe_B(T)}{\delta} \cdot \left(p_{B,ret}^{0.5} - p_{B,perm}^{0.5}\right) \tag{1.22}$$

where δ is the thickness of the dense membrane, $p_{B,ret}$ and $p_{B,perm}$ are the partial pressures of B in the retentate and permeate regions, respectively and:

$$Pe_B(T) = Pe_B^0 \cdot \exp\left(-\frac{E_a}{R \cdot T}\right) \tag{1.23}$$

is the permeability which depends on the temperature T according to the Arrhenius behavior.

Pe_B^0 is the pre-exponential factor, and E_a is the apparent activation energy. The partial pressure in the permeate side can be controlled by means of the total pressure and by the flow rate of the sweeping gas.

However, in order to allow a given flow rate of a product to be removed from the reaction mixture a suitable membrane surface area must be available.

Once flow rate W_B and the other operating parameters are given, the calculation of the membrane surface area S is straightforward since is:

$$S = \frac{W_B}{J_B} \tag{1.24}$$

When $p_{B,ret}$ and $p_{B,perm}$ are constant at each point of the membrane, the calculation is very simple. This could be the case of a membrane module located outside the reactor.

However, in the case of integrated MBRs, with the membrane inside the reactor, partial pressures $p_{B,ret}$ and $p_{B,perm}$ and, often, temperature T change, in general, along the tubular reactor. Therefore, Eq. 1.22 gives the local flux at a given section of the MBR, and Eq. 1.24 must be given in differential form:

$$dS = \frac{dW_B}{J_B} \tag{1.25}$$

where dW_B indicates the infinitesimal flow rate of B crossing the membrane through the surface dS. The overall surface area is, then, calculated by integration of Eq. 1.25.

Anyway, the surface area of a given membrane depends on the flow rate, the thickness of the layer, the pressure difference, and the temperature. Compatibly with specific requirements of the process, required membrane surface can be minimized through a suitable choice of the operating parameters, especially of the pressures. This is an important issue due to the generally high specific cost of selective membranes.

1.7 Final Remarks

The potential of MRs is enormous. In particular, they play a key role in reversible endothermic reactions where their use can provide high reactant conversion at relatively lower temperatures. This will make possible to avoid fuel combustion in fired heaters to supply the required reaction heat and could allow multiple heat sources at low temperature to be used, ranging from gas turbine exhausts to solar heated molten salts or even helium from a nuclear power [24, 25, 29].

A new configuration will then emerge where heat and power are cogenerated and a considerable energy saving is achieved, reducing the products' manufacturing costs. The cogenerative scheme presented in Fig. 1.8 refers to natural gas steam reforming, but this concept can easily be extended to other chemical processes.

The number of possible applications of MRs is indeed large, but commercial applications are emerging slowly due to a number of practical issues to be solved such as membrane stability, mass transfer limitation (low flux), high membrane production costs etc. This is still a task far from completion and will require a close cooperation of catalyst scientists, material scientists, and chemical engineers. Chap. 2 was then devoted to the membranes state of the art to better understand what kind of progress is required in terms of materials and engineering problems. Cost of manufacture could be considerably reduced if membranes will find bulk applications; Chap. 3 of this book deals with technical solutions aimed to such cost reduction. Chap. 4 is devoted to mathematical modeling which is a tool widely used in process development and optimisation.

A part from the steam reforming there are other interesting applications in the petrochemical industry: methane dry reforming, catalytic partial oxidation, autothermal reforming, water gas shift, H_2S cracking, and hydrocarbon dehydrogenation. More details about some of these applications are given in Chaps. 5, 6, 7, 8, and 9.

Finally, in Chap. 10 the RMM pilot plant shown in Fig. 1.4 is described, and some preliminary operational data are presented and discussed.

References

1. Sanchez Marcano JG, Tsotsis TT (2002) Catalytic membranes and membrane reactors. Wiley, Weinheim
2. Drioli E (2004) Membrane reactors. Chem Eng Proc 43:1101–1102
3. Mardilovich PP, She Y, Rei MH, Ma YH (1998) Defect-free palladium membranes on porous stainless-steel support. AIChE J 44:310
4. Li A, Liang W, Ronald H (1998) Characterisation and permeation of palladium/stainless steel composite membranes. J Membr Sci 149:259–268
5. Paglieri SN, Foo KY, Collins JP, Harper-Nixon DL (1999) A new preparation technique for Pd/alumina membranes with enhanced high-temperature stability. Ind Eng Chem Res 38:1925–1936

6. McCool B, Xomeritakis G, Lin YS (1999) Composition control and hydrogen permeation characteristics of sputter deposited palladium–silver membranes. J Membr Sci 161:67–76
7. Shu J, Grandjean B, Kaliaguine S (1994) Methane steam reforming in asymmetric Pd and Pd-Ag porous SS membrane reactors. Appl Catal A Gen 119:305–325
8. Lin Y, Liu S, Chuang C, Chu Y (2003) Effect of incipient removal of hydrogen through palladium membrane on the conversion of methane steam reforming: experimental and modeling. Catal Today 82:127–139
9. Madia G, Barbieri G, Drioli E (1999) Theoretical and experimental analysis of methane steam reforming in a membrane reactor. Can J Chem Eng 77:698–706
10. Chai M, Machida M, Eguchi K, Arai H (1994) Promotion of hydrogen permeation on a metal-dispersed alumina membrane and its application to a membrane reactor for steam reforming. Appl Catal A Gen 110:239–250
11. Gallucci F, Paturzo L, Basile A (2004) A simulation study of steam reforming of methane in a dense tubular membrane reactor. Int J Hydrogen Energy 29:611–617
12. Itoh N, Shindo Y, Haraya H, Hakuta T (1988) A membrane reactor using microporous glass for shifting equilibrium of cyclohexane dehydrogenation. J Chem Eng Jpn 21:399–404
13. Wu JCS, Gerdes TE, Pszczolkowski JL, Bhave RR, Liu PKT (1990) Dehydrogenation of ethylbenzene to styrene using commercial ceramic membranes as reactors. Sep Sci Technol 25:1489–1510
14. Becker YL, Dixon AG, Moser WR, Ma YH (1993) Modelling of ethylbenzene dehydrogenation in a catalytic membrane reactor. J Membr Sci 77:233–244
15. Basile A, Drioli E, Santella F, Violante V, Capannelli G, Vitulli G (1995) A study on catalytic membrane reactors for water gas shift reaction. Gas Sep Purif 10:53
16. Basile A, Criscuoli A, Santella F, Drioli E (1996) Membrane reactor for water gas shift reaction. Gas Sep Purif 10:243
17. Criscuoli A, Basile A, Drioli E (2000) An analysis of the performance of membrane reactors for the water-gas shift reaction using gas feed mixtures. Catal Today 56:53
18. Basile A, Chiappetta G, Tosti S, Violante V (2001) Experimental and simulation of both Pd and Pd/Ag for a water gas shift membrane reactor. Sep Purif Technol 25:549
19. Iyoha O, Enick R, Killmeyer R, Howard B, Morreale B, Ciocco M (2007) Wall-catalyzed water-gas shift reaction in multi-tubular Pd, 80wt%Pd-20 wt%Cu membrane reactors at 1173 k. J Membr Sci 298:14
20. Brunetti A, Barbieri G, Drioli E, Granato T, Lee K-H (2007) A porous stainless steel supported silica membrane for WGS reaction in a catalytic membrane reactor. Chem Eng Sci 62:5621
21. Giessler S, Jordan K, da Diniz Costa JC, Lu GQM (2003) Performance of hydrophobic and hydrophilic silica membrane reactors for the water gas shift reaction. Sep Purif Technol 33:255
22. Battersby S, Duke MC, Liu S, Rudolph V, da Diniz Costa JC (2008) Metal doped silica membrane reactor: operational effects of reaction and permeation for the water gas shift reaction. J Membr Sci 316:46
23. Akamatsu K, Nakane M, Sugawara T, Hattori T, Nakao S (2008) Development of a membrane reactor for decomposing hydrogen sulphide into hydrogen using a high-performance amorphous silica membrane. J Membr Sci 325:16
24. Galuszka J, Iaquaniello G (2009) Membrane assisted conversion of Hydrogen sulphide-Patent International Application, PCT/CA2009/001562 filed on October 29, 2009
25. De Falco M, Barba D, Cosenza S, Iaquaniello G, Farace A, Giacobbe FG (2009) Reformer and membrane modules plant to optimize natural gas conversion to hydrogen, Special Issue of Asia-Pacific J Chem Eng Mem React. doi:10.1002/apj.241
26. De Falco M, Barba D, Cosenza S, Iaquaniello G, Marrelli L (2008) Reformer and membrane modules plant powered by a nuclear reactor or by a solar heated molten salts: assessment of the design variables and production cost evaluation. Int J Hydrogen Energy 33:5326–5334

27. Caravella A, Di Maio FP, Di Renzo A (2010) Computational study of staged membrane reactor configurations for methane steam reforming. I. Optimization of stage lengths. AIChE J 56(1):248–258
28. Caravella A, Di Maio FP, Di Renzo A (2010) Computational study of staged membrane reactor configurations for methane steam reforming. II. Effect of number of stages and catalyst amount. AIChE J 56(1):259–267
29. Barba D, Giacobbe F, De Cesaris A, Farace A, Iaquaniello G, Pipino A (2008) Membrane reforming in converting natural gas to hydrogen (part one). Int J Hydrogen Energy 33:3700–3709

Chapter 2
Pd-based Selective Membrane State-of-the-Art

A. Basile, A. Iulianelli, T. Longo, S. Liguori and Marcello De Falco

Abbreviations

AASR	Acetic acid steam reforming
BESR	Bioethanol steam reforming
CVD	Chemical vapour deposition
ELP	Electroless plating deposition
EP	Electroplating
ESR	Ethanol steam reforming
EVD	Electrochemical vapour deposition
FBR	Fixed bed reactor
GSR	Glycerol steam reforming
HTR	High temperature reactor
IUPAC	International Union of Pure and Applied Chemistry
LTR	Low temperature reactor
ML	Molecular layering
MR	Membrane reactor
MS	Magnetron sputtering
MSR	Methane steam reforming
PEMFC	Proton exchange membrane fuel cell
POM	Partial oxidation of methane
PSA	Pressure swing adsorption
PVD	Physical vapour deposition
SRM	Methanol steam reforming

A. Basile (✉), A. Iulianelli, T. Longo and S. Liguori
Institute on Membrane Technology of National Research Council (ITM-CNR),
Via P. Bucci Cubo 17/C c/o University of Calabria, 87036 Rende, CS, Italy
e-mail: a.basile@itm.cnr.it

M. De Falco
Faculty of Engineering, University Campus Bio-Medico of Rome,
via Alvaro del Portillo 21, 00128 Rome, Italy

M. De Falco et al. (eds.), *Membrane Reactors for Hydrogen Production Processes*,
DOI: 10.1007/978-0-85729-151-6_2, © Springer-Verlag London Limited 2011

WGS	Water gas shift
WHSV	Weight hourly space velocity

List of Symbols

D	Diffusion coefficient
d_p	Pore diameter
E_a	Apparent activation energy
G	Geometrical factor
J	Flux or permeation rate
$J_{H_2,Sieverts-Fick}$	Hydrogen flux through the membrane according to Sieverts–Fick law
J_{H_2}	Hydrogen flux through the membrane
J_i	Flux of the i-species across the membrane
J_m	Mass flux
M_i	Molecular weight of the i-species
n	Dependence factor of the hydrogen flux on the hydrogen partial pressure
p	Pressure
$Pe^0_{H_2}$	The pre-exponential factor
Pe_{H_2}	The hydrogen permeability
$p_{H_2,perm}$	Hydrogen partial pressures at the permeate side
$p_{H_2,ret}$	Hydrogen partial pressures at the retentate side
R	Universal gas constant
T	Absolute temperature
X	Coordinate perpendicular to the transport barrier
$\Delta H^\circ_{298\,K}$	Enthalpy variation in standard conditions
Δp_i	Pressure difference of species
α	Ideal separation factor or selectivity
δ	Membrane thickness
ϕ_{pore}	Pore diameter

2.1 Introduction

The first scientific study on palladium-based membranes, available in the Elsevier Scopus database [1], where more than 6,000 scientific journals are taken into account, is dated 1955, when Juenker et al. [2] analyzed the use of palladium membranes for hydrogen purification. Today, it is well known that the palladium membranes are, mainly, applied in the field of gas separation and, particularly, in the issue of the hydrogen rich-stream purification [3]. As reflected by the data of Fig. 2.1, the scientific interest towards palladium-based membranes is increased

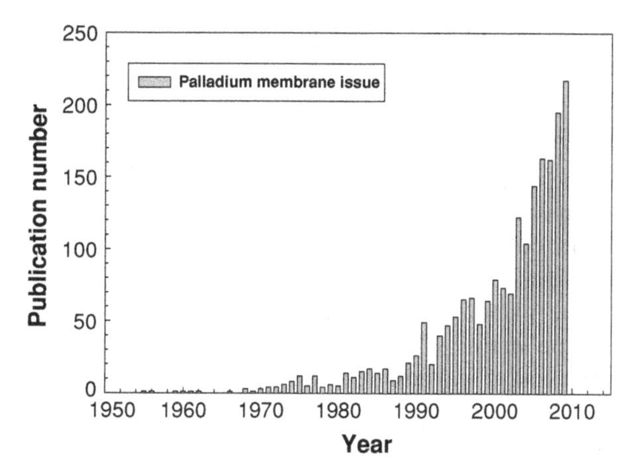

Fig. 2.1 Number of published papers per year on palladium membrane applications

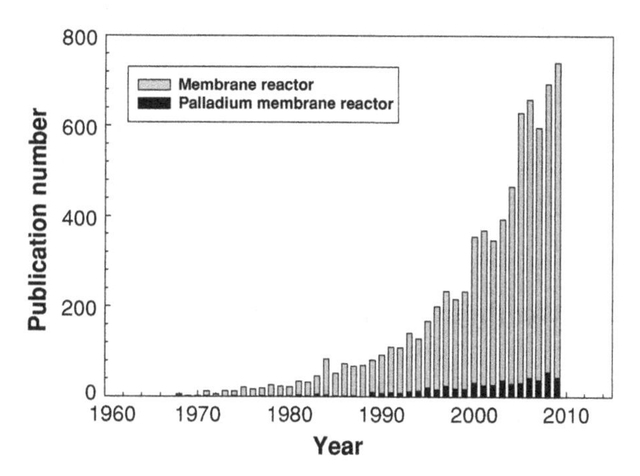

Fig. 2.2 Number of publications per year on membrane reactors area and on restricted area on palladium-based membrane reactors

especially in the last two decades. The statistics on the scientific publications in the contest of palladium membranes applications reported in the figure were made by means of Elsevier Scopus database.

Moreover, Fig. 2.2 points out further statistics data on palladium membranes applied in the field of membrane reactors (MRs), devices combining the separation properties of the membranes with the typical characteristics of catalytic reaction steps in only one unit. In particular, this figure reports the number of publications on palladium-based membranes reactors with respect to the total number of publications in the membrane reactors area.

The progress in the field of palladium-based MRs is due to their capacity to produce a pure hydrogen stream, owing to infinite hydrogen perm-selectivity with respect to all other gases. Moreover, in the last years the "hydrogen economy" has taken place in order to solve the problematic concerning the climate change and air pollution due to the emissions caused by the use of fossil fuels [4]. In particular, the "hydrogen economy" has been developed with the aim of using hydrogen as an energy carrier, producible from renewable sources as an alternative to fossil fuels.

Nevertheless, the commercialization of pure palladium membranes is still limited by several factors:

- pure palladium membranes undergo the embrittlement phenomenon when exposed to pure hydrogen at temperatures below 300°C,
- pure palladium membranes are subject to deactivation by carbon compounds at temperature above 450°C,
- pure palladium membranes are subject to irreversible poisoning by sulphur compounds,
- the cost of palladium is high.

In order to reduce the aforementioned drawbacks, palladium can be alloyed with a variety of other metals in order to manufacture membranes able to increase the hydrogen permeability shown by the pure palladium membranes.

As a main scope, the present chapter will give an overview on the general classification of the membranes, paying particular attention to the palladium-based membranes and their applications, pointing out the most important benefits and the drawbacks due to their use. Finally, the application of palladium-based membranes in the area of the membrane reactors will be illustrated and such reaction processes in the issue of hydrogen production will be discussed.

2.2 Membrane Classification

As indicated by IUPAC definition [5], a *membrane* can be described as a structure having lateral dimensions much greater than its thickness through which mass transfer may occur under a variety of driving forces such as gradient of concentration, pressure, temperature, electric potential, etc. A schematic representation of a two-phase system separated by a membrane is given in Fig. 2.3, where the Phase 1 is usually considered as the feed, while the Phase 2 as the permeate.

As schematically resumed in Fig. 2.4, the membranes are classified on the base of their nature, geometry and separation regime [6].

The classification based on the membrane nature distinguishes them into biological and synthetic ones, differing completely for functionality and structure [7].

Biological membranes are easy to be manufactured, but they present many drawbacks such as limited operating temperature (below 100°C) and pH range, problems related to the clean-up and susceptibility to microbial attack due to their natural origin [7].

Synthetic membranes can be subdivided into organic (polymeric) and inorganic (ceramic, metallic) ones according to their operative temperature limit: *polymeric membranes* commonly operate between 100 and 300°C [8], above 200°C the *inorganic ones*.

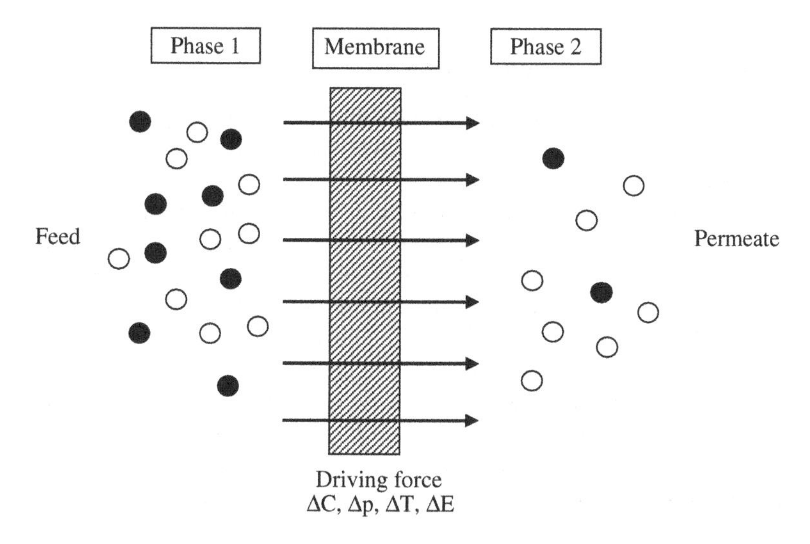

Fig. 2.3 Schematic representation of a two-phase system separated by a membrane

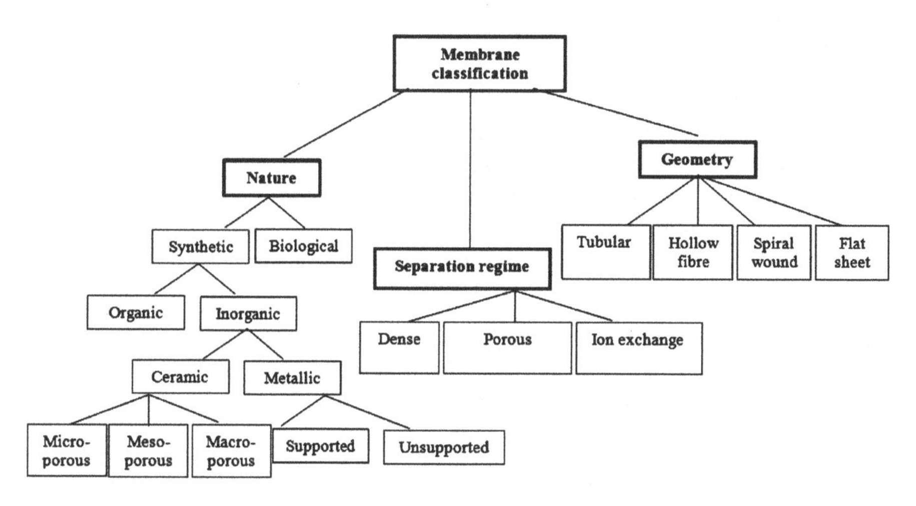

Fig. 2.4 Scheme of a general classification of the membranes

In the viewpoint of the morphology and/or membrane structure categorization, the inorganic membranes can be also subdivided into ceramics and metallic. In particular, *ceramics membranes* differ according to their pore diameter in microporous ($d_p < 2$ nm), mesoporous (2 nm $< d_p <$ 50 nm) and macroporous ($d_p >$ 50 nm) [5]. Finally, *metallic membranes* can be categorized into supported and unsupported.

Generally, inorganic membranes are stable between 200 and 800°C and in some cases they can operate at elevated temperatures (ceramic membranes over 1000°C) [9].

Depending on their geometry, the membranes can be subdivided in *tubular, hollow fibre, spiral wound* and *flat sheet* [10]:

- *Tubular membranes* are easy to clean and show good hydrodynamic control, but as important drawbacks they require relatively high volume per membrane area unit and present high costs.
- *Hollow fibre membranes* can be considered as practical and cheaper alternatives than conventional chemical and physical separation processes. They offer high packing densities and they can withstand relatively high pressure owing to their structural integrity. In this contest, they allow flexibility in system design and operation.
- *Spiral wound membranes* offer advantages such as compactness, good membrane surface/volume and low capital/operating cost ratios. Nevertheless, they are not suitable for viscous fluid and are difficult to clean.
- *Flat sheet membranes* offer moderate membrane surface/volume ratios. However, they are susceptible to plugging due to flow stagnation points, difficult to clean and expensive.

A further membrane classification is based on the separation mechanism. There are three separation mechanisms depending on specific properties of the components [11]:

1. separation based on molecules/membrane surface interactions (e.g., multi-layer diffusion) and/or difference between the average pore diameter and the average free path of fluid molecules (e.g. *Knudsen mechanism*);
2. separation based on the difference of diffusivity and solubility of substances in the membrane: *solution/diffusion mechanism*;
3. separation due to the difference in charge of the species to be separated: *electrochemical effect.*

Based on these mechanisms, the membranes can be further classified *in porous, dense* and *ion-exchange*. In Table 2.1, the different diffusion mechanisms for porous and dense membranes are reported.

In the case of porous membranes:

- *Poiseuille (viscous flow) mechanism* occurs when the average pore diameter is bigger than the average free path of fluid molecules. In this case, the collisions among the different molecules are more frequent than those among the molecules and the porous wall: as a consequence no separation takes place [12].

Table 2.1 Diffusion mechanisms in porous and dense membranes

Membrane	d_{pore} (nm)	Diffusion mechanism
Macroporous	>50	Poiseuille (Viscous flow)
Mesoporous	2–50	Knudsen
Microporous	<2	Activated process
Dense Pd	–	Fick

- *Knudsen mechanism* occurs when the average pore diameter is similar to the average free path of fluid molecules. In this case, the collisions of the molecules with the porous wall are very frequent and the flux of the component permeating through the membrane is calculated by means of the following equation [12]:

$$J_i = \frac{G}{\sqrt{2 \cdot M_i \cdot R \cdot T}} \cdot \frac{\Delta p_i}{\delta} \tag{2.1}$$

  where J_i is the flux of the i-species across the membrane, G the geometrical factor, which takes into account the membrane porosity and the pore tortuosity, M_i molecular weight of the i-species, R universal gas constant, T absolute temperature, Δp_i pressure difference of species and δ membrane thickness.
- *Surface diffusion* is achieved when one of the permeating molecules is adsorbed on the pore wall due to the active sites present in the membrane [13]. This type of mechanism can reduce the effective pore dimensions, unfavouring the transfer of different molecular species [14]. However, this diffusion can take place also in the presence of a Knudsen transport. This mechanism is less significant by increasing the temperature owing to the progressive decrease of the bond strength between molecules and surface.
- *Capillary condensation* occurs when one of the components condenses within the pores due to capillary forces, which are sufficiently strong only at low temperature and in presence of small pores. If the pores dimension is small and homogeneous and the pores are uniformly distributed on the membrane, this mechanism can offer high selectivity [15, 16]. Generally, the capillary condensation favours the transfer of relatively large molecules [17].
- *Multi-layer diffusion* is developed when the molecule/surface interactions are strong. This mechanism is like to an intermediate flow regime between surface diffusion and capillary condensation [18].
- *Molecular sieve* takes place when the pore diameters are very small, allowing the permeation of only smaller molecules.

In the case of dense membranes, the transport mechanism is a *solution–diffusion* mechanism, in which the dissociated molecules on the gas/membrane interface are adsorbed at the atomic level on the membrane surface. The atoms diffuse through the membrane and are re-combined to form molecules at the gas/membrane interface. Afterwards, they desorb.

Among all types of membranes, the inorganic dense membranes have attracted the interest of many researchers due to their capacity to separate completely a product from gaseous mixtures [19]. In particular, the dense palladium membranes are used because of their complete hydrogen perm-selectivity. In the last years, the increasing interest towards this type of membranes is, also, due to hydrogen application as energy carrier.

For this reason, in the following, a general introduction to pure palladium membranes is given.

2.3 Hydrogen Production and Palladium Membranes

In the current fossil fuel economy, the fossil fuels burning causes the emission of greenhouse gases and other pollutants. In order to mitigate the air pollution and the climate change, the use of alternative technologies is become necessary. In this contest, great interest is paid to PEMFCs, which are capable to produce electricity directly from hydrogen and oxygen, without combustion, making the process non-polluting [20]. A PEMFC uses a permeable polymeric membrane as the electrolyte (Fig. 2.5).

PEMFCs are characterized by low operative temperature (80–100°C), high current density, compactness, fast start-up and suitability for discontinuous operation [21]. These features make PEMFCs the most promising and attractive candidate for a wide variety of power applications ranging from portable/micropower and transport to large-scale stationary power systems for buildings and distributed generation [22], as shown in Fig. 2.6.

Today, the hydrogen for supplying the PEMFCs derives, mainly, from fossil fuels (48% from natural gas, 30% from oil and 18% from coal) [24]. Nevertheless, owing to the climate change as well as the cost increase of oil and gas, the development of a strategy for exploiting alternative and renewable sources represent a top priority in which hydrogen could be an inexhaustible energy carrier [25].

Nevertheless, the full commercialization of PEMFC systems needs a stable supply of hydrogen, which must be characterized by high purity for avoiding the CO poisoning of the PEMFC anodic catalyst [26].

Nowadays, the dominant technology for direct hydrogen production is steam reforming from hydrocarbons. Generally, reforming reactions, carried out in conventional fixed bed reactors (FBRs), produce a hydrogen rich gas mixture containing carbon oxides and other by-products as well as the unreacted reactants. Therefore, in the viewpoint of feeding a PEMFC, which can tolerate only

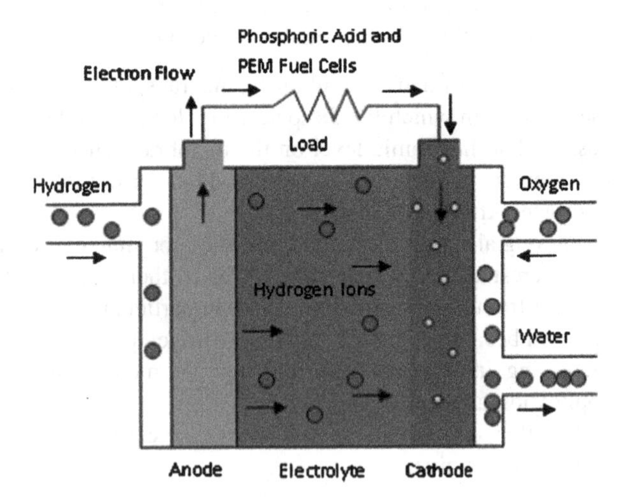

Fig. 2.5 Diagram of a PEM fuel cell

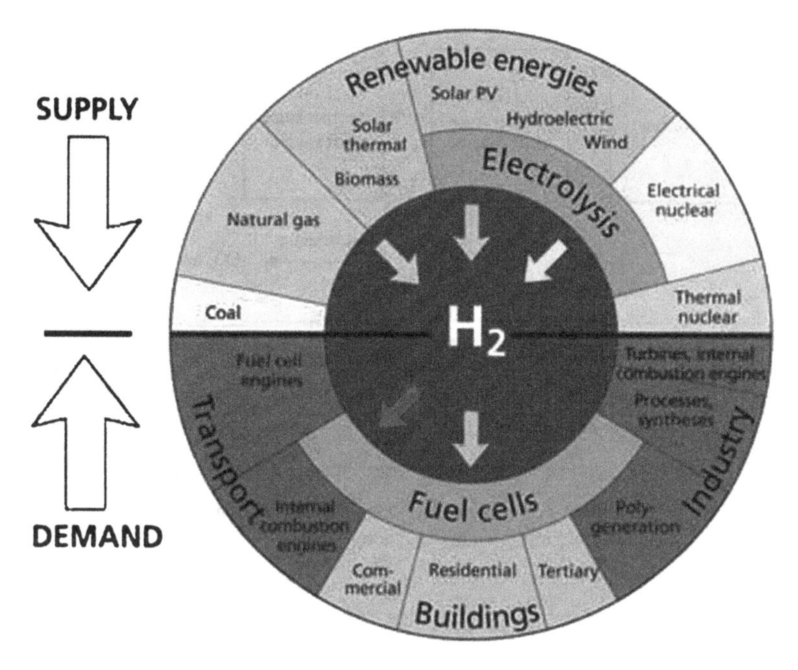

Fig. 2.6 Summary of the hydrogen economy. *Upper part* production, *lower part* uses [23]

few ppm of CO, the hydrogen going out from a conventional reformer needs to be purified by means of the following processes: two-steps water gas shift reactors followed by a separation/purification unit (PSA, Pd membrane, etc.), as reported in Fig. 2.7a [27].

Many researchers have proposed, as an economically more advantageous method, the use of a process able to produce a pure hydrogen stream in only one system [28–33]. In this contest, dense palladium MRs are able to carry out both the reaction and separate pure hydrogen in the same device (Fig. 2.7b).

In fact, although, niobium (Nb), vanadium (V) and tantalum (Ta) offer higher hydrogen permeability than palladium in a temperature range between 0 and 700°C, as shown in Fig. 2.8, nevertheless these metals give a stronger surface resistance to hydrogen transport than the palladium (Pd). For this reason, dense palladium membranes are preferentially used.

The hydrogen molecular transport in the palladium membranes occurs through a solution/diffusion mechanism, which follows six different activated steps [35]:

- dissociation of molecular hydrogen at the gas/metal interface;
- adsorption of the atomic hydrogen on membrane surface;
- dissolution of atomic hydrogen into the palladium matrix;
- diffusion of atomic hydrogen through the membrane;
- re-combination of atomic hydrogen to form hydrogen molecules at the gas/metal interface;
- desorption of hydrogen molecules.

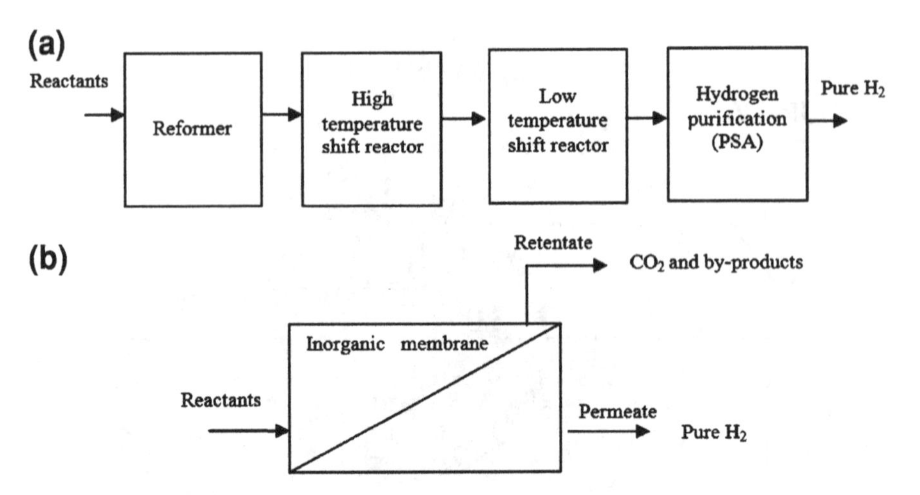

Fig. 2.7 Scheme of pure hydrogen production by hydrocarbons compounds steam reforming: traditional scheme (**a**) and inorganic membrane reactor (**b**) [28]

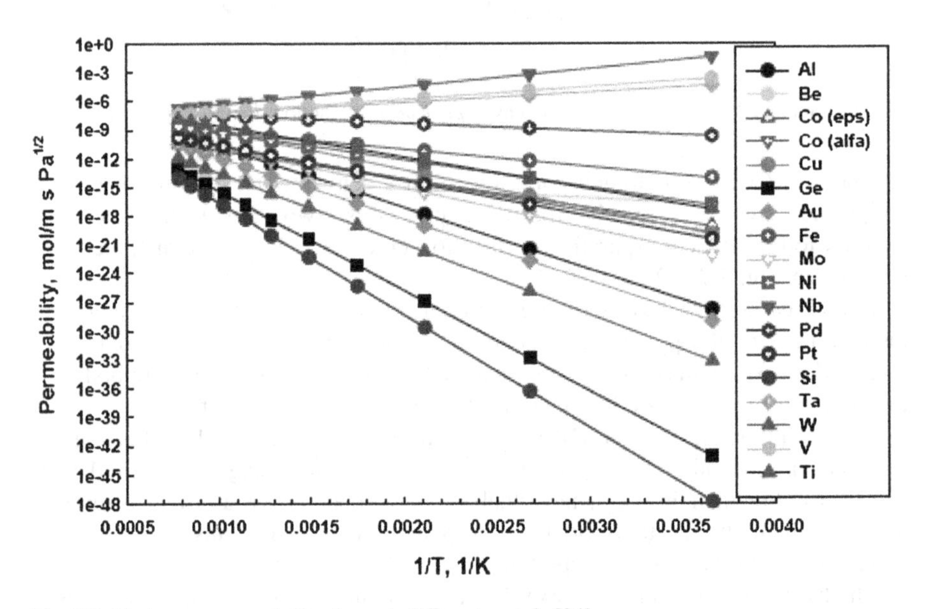

Fig. 2.8 Hydrogen permeability through different metals [34]

Depending on temperature, pressure, gas mixture composition and thickness of the membrane, each one of these steps may control hydrogen permeation through the dense film [28]. As a result, the hydrogen permeating flux can be expressed by means of the following equation:

$$J_{H_2} = Pe_{H_2}/\delta \cdot \left(p^n_{H_2,ret} - p^n_{H_2,perm}\right) \tag{2.2}$$

where n (variable in the range 0.5–1) is the dependence factor of the hydrogen flux to the hydrogen partial pressure, J_{H_2} the hydrogen flux permeating through the membrane, Pe the hydrogen permeability, δ the membrane thickness, $p_{H_2,ret}$ and $p_{H_2,perm}$ the hydrogen partial pressures in the retentate (the reaction side) and permeate sides (the volume in which the hydrogen permeating through the membrane is collected), respectively.

This equation even points out the inverse proportionality to the membrane thickness. The role of the membrane thickness is very important: on one hand, a thinner membrane offers a higher permeability; on the other hand, thicker membranes are necessary in order to ensure the mechanical resistance and strength.

When the pressure is relatively low, the diffusion is assumed to be the rate-limiting step and the factor n is equal to 0.5. In this case, Eq. 2.2 becomes the Sieverts–Fick law [36]:

$$J_{H_2,\text{Sieverts-Fick}} = Pe_{H_2}/\delta \cdot \left(p_{H_2,ret}^{0.5} - p_{H_2,perm}^{0.5}\right) \tag{2.3}$$

On the contrary, at high pressures the hydrogen–hydrogen interactions in the palladium bulk are not negligible. In this case, n becomes equal to 1:

$$J_{H_2} = Pe_{H_2}/\delta \cdot \left(p_{H_2,ret} - p_{H_2,perm}\right) \tag{2.4}$$

The relationship between hydrogen permeability and temperature follows an Arrhenius behaviour (Eq. 2.5), while "n" generally does not depend on the temperature:

$$Pe_{H_2} = Pe_{H_2}^{0}\exp(-E_a/RT) \tag{2.5}$$

where Pe_0 is the pre-exponential factor, E_a the apparent activation energy, R the universal gas constant and T the absolute temperature.

As a consequence, when Sieverts–Fick law is valid, the hydrogen flux is written in terms of the so-called Richardson's equation:

$$J_{H_2} = Pe_{H_2}^{0}[\exp(-E_a/RT)]\left(p_{H_2,ret}^{0.5} - p_{H_2,perm}^{0.5}\right)/\delta \tag{2.6}$$

Nevertheless, although the pure palladium membranes are characterized by a complete hydrogen perm-selectivity, their commercialization is limited by some drawbacks such as relatively low hydrogen permeability and high cost [37].

2.3.1 Problems Associated with the Pure Palladium Membranes

The most important problem associated with the use of pure palladium membranes is the "hydrogen embrittlement" phenomenon. When the temperature is below 300°C and the pressure below 2.0 MPa, the β-hydride phase may nucleate from the

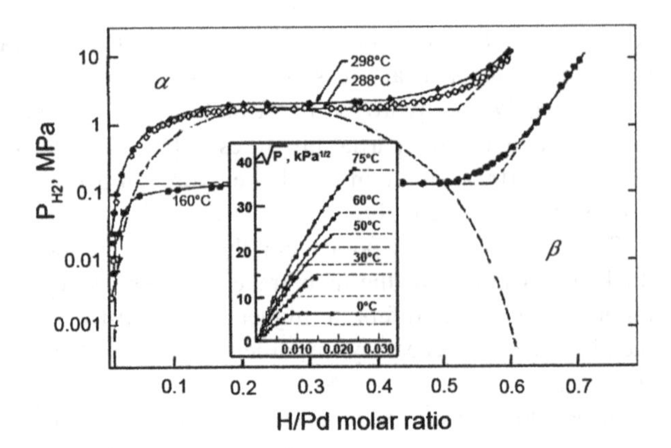

Fig. 2.9 Equilibrium solubility isotherms of PdH_n for bulk Pd at different temperatures [42]

α-phase, resulting in severe lattice strains (see Fig. 2.9), so that a pure palladium membrane becomes brittle after a few cycles of $\alpha \leftrightarrows \beta$ transitions [38–40].

These transitions do not take place as a change of the lattice structure, but as a lattice dilatation. The β-hydride phase formation is represented as a clustering of hydrogen atoms, whose energy of attraction, being associated with the lattice, strains around the dissolved hydrogen atom [41].

A possible solution to avoid this phenomenon is represented by the use of a Pd-alloy containing another metal, such as silver. The role of silver is explained by its electron donating behaviour, being largely similar to the one of the hydrogen atom in palladium. Silver and hydrogen atoms would compete for the filling of electron holes [42].

Another critical problem is represented by the palladium surface contamination of Hg vapour, hydrogen sulphide, SO_2, thiophene, arsenic, unsaturated hydrocarbons, chlorine carbon from organic materials. In particular:

- *Poisoning of sulphur compounds*: Pd-coated membranes could rapidly be destroyed after exposure to a gas stream containing hydrogen sulphide and the poisoning effects are irreversible [43, 44]. Palladium becomes palladium sulphide, whose lattice constant is twice than of pure Pd and, thus, the structural stress leads to the formation of cracks.
- *Poisoning of CO*: the presence of CO in a feed gas stream could cause a decrease in the hydrogen permeation flux, because the adsorbed CO displaces the adsorbed hydrogen and further blocks hydrogen adsorption sites [45]. Moreover, this reduction becomes more significant at low temperature (below 150°C) or high CO concentration [46]. CO is adsorbed on the palladium surface blocking available dissociation sites for hydrogen. In order to improve the chemical stability of the metal membranes it is possible:
  - to use different types of Pd alloys constituted by other metals such as Cu, Ni, Fe, Pt and Ag,
  - to prepare nanostructured or amorphous thin alloy membranes.

- *Poisoning of H_2O*: the presence of water vapour has a more negative effect on hydrogen permeability than the presence of CO [47]. The adsorbed water molecules dissociate on the surface of the Pd film:

$$H_2O_{ads} \rightarrow OH_{ads} + H_{ads} \quad \text{and/or} \quad H_2O_{ads} \rightarrow O_{ads} + 2H_{ads} \tag{2.7}$$

  where the H_{ads} may permeate into the bulk of the Pd film [48]. H_2O is recombined through these reactions:

$$2OH_{ads} \rightarrow H_2O_{gas} + O_{ads} \quad \text{and/or} \quad O_{ads} + 2H_{ads} \rightarrow H_2O_{gas} \tag{2.8}$$

  Therefore, the process of H_2O dissociation/recombinative desorption contaminates the palladium surface with adsorbed O.
- *Poisoning of coke*: both hydrogen permeance and perm-selectivity of a thin palladium membrane decrease after it is brought in contact with coke at elevated temperature [49]. This phenomenon can be addressed for the fact that carbon atoms penetrate into the palladium lattice and cause the failure of the membrane owing to the expansion of the palladium lattice.

Moreover, the palladium membranes are still very expensive. In order to reduce their cost it is possible to develop low palladium content based alloy trying to reduce the Pd-based layer thickness [50–52].

2.4 Palladium-based Membranes

Generally, the palladium-based membranes can be subdivided into supported and laminated ones. In the *supported membranes*, a thin dense layer of a palladium alloy is deposited onto a porous support such as porous Vycor glass (silica gel). Nevertheless, using this kind of support, the palladium layer is easily stripped off owing to the loss of an anchor effect [53].

Other types of porous glass materials are represented by SiO_2, Al_2O_3 and B_2O_3, giving excellent anchor effect and adherence [53]. Also porous stainless steel (PSS) can be considered as a valid support due to its mechanical durability, its thermal expansion coefficient close to that of palladium and the ease of gas sealing [53].

The upper temperature limits of the supported membranes depend on: the material, the chemical atmosphere and the support characteristics such as porosity and pore diameter [54].

In the *laminated membranes*, thin palladium (or palladium alloys) layer avoids the formation of oxides on the metallic surfaces resulting in a reduction of the hydrogen adsorption activation energy and, as a consequence, in an increase of the hydrogen permeation flux [55].

Generally, the palladium alloys have some advantages with respect to the pure palladium membranes such as a reduced critical temperature for the α–β phase transition. For example, Pd–Ag membranes can operate in hydrogen atmosphere at

Table 2.2 Improvement in hydrogen permeability of various binary and tertiary palladium alloys at 350°C [57]

Alloy metal	wt% for maximum permeability	Normalized permeability (Pe_{alloy}/Pe_{Pd})
Y	10.0	3.8
Ag	23.0	1.7
Ce	7.7	1.6
Cu	40.0	1.1
Au	5.0	1.1
Ru, In	0.5–6.0	2.8
Ag, Ru	30.2	2.2
Ag, Rh	19.1	2.6
Pure Pd	–	1.0

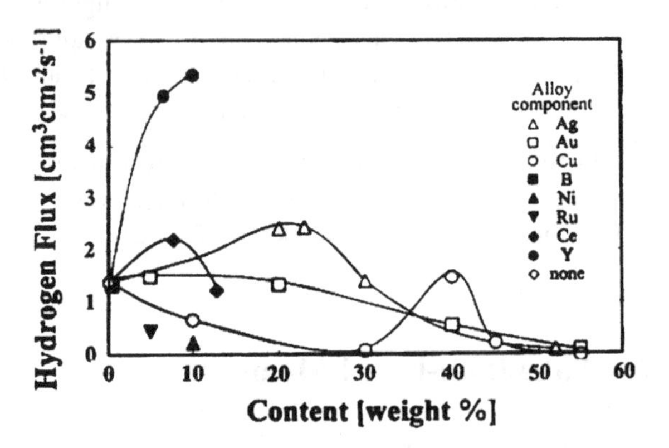

Fig. 2.10 Hydrogen flux through palladium alloy membranes against metal content [58]

temperatures below 300°C without observing embrittlement rather than pure palladium membranes [42]. Moreover, in some cases, the hydrogen permeability of palladium alloys is higher than pure palladium, as reported in Table 2.2. In fact, as shown in Fig. 2.10, the hydrogen flux through the Pd–Ag membranes reaches the maximum value at 350°C and 2.2 MPa with a 23% Ag content. In details, the permeability is 1.7 times higher than one of a pure Pd membrane. The Pd–Cu alloy even shows a maximum value of hydrogen flux with 40% Cu content, although these membranes suffer a permeation decrease when exposed at 900°C for a long time [56].

Moreover, the palladium alloys improve chemical resistance of the membranes. For example, Pd–Cu and Pd–Au increase the resistance to H_2S [59] as well as palladium-coated amorphous Zr–M–Ni (M = Ti, Hf) alloy membranes are resistant enough in a hydrogen atmosphere and have stable hydrogen permeability in the range of 200–300°C [60].

In order to further reduce the membrane cost, low palladium content based alloys can be produced [52]. In fact, Basile et al. [61] demonstrated that thin dense Ti–Ni–Pd membranes with a low palladium content (4.17 vs 77.0% of the Pd–Ag and 60.0% of the Pd–Cu) make this membrane still competitive from an economical point of view.

2.4.1 Methods for Producing Palladium-based Membranes

Palladium-based membranes can be produced by several methods, depending on some factors such as the nature of the metal itself, the manufacturing facilities, required thickness, surface area, shape, purity, etc. Nevertheless, no one method can produce a membrane, which combines advantageously all these factors. Therefore, the choice of the production method becomes a compromise between these factors [42]. At lab-scale, the thickness as well as the continuity and imperviousness of the film are considered the most important factors [42]. In particular, in the last few years, the main aim has been to reduce the thickness of the palladium-based films.

Criscuoli et al. [62], for example, carried out an economical analysis of palladium-based MRs, comparing the costs between the conventional apparatus and different membrane systems for pure hydrogen production. In particular, the authors studied the effect of the palladium thickness and membrane permeability to hydrogen on the costs of membrane devices. They demonstrated that both higher permeabilities and lower thickness improve the hydrogen removal leading to a decrease in membrane area for a pure hydrogen recovery and a cost reduction. The authors concluded that MRs could represent a possible alternative to conventional apparatus for palladium thicknesses equal or lower than 20 μm. Moreover, the authors highlighted that current methods of preparing palladium-based membranes are still expensive due to the inadequate productivity per unit membrane area. A cost reduction for the Pd-based membranes could be achieved through a massive production as detailed in Chap. 3.

The most important production methods of palladium-based membranes are described in the following.

The *Conventional cold rolling* is the most diffuse technique for producing metallic plates or sheets at laboratory scale [63]. It involves:

- melting the raw materials with chosen composition at very high temperature,
- ingot casting,
- high temperature homogenization,
- hot and cold forging or pressing, followed by repeated sequences of alternate cold rolling and anneals, down to the required thickness.

If the cooling speed of melts is fast, amorphous materials (*metallic glasses*) can be realized obtaining good characteristics such as high mechanical toughness, considerable corrosion resistance, good electronic properties, high catalytic activities, reversible hydrogen storage, etc. [64, 65]. The cold rolling treatment can enhance the hydrogen solubility in palladium and its alloys owing to the accumulation of hydrogen excess in the stress field around dislocations, formed during the process. This effect can be gradually eliminated during annealing of the deformed membranes by increasing the temperature [66].

In *Physical vapour deposition* (PVD) method, the solid material to be deposited is evaporated in a vacuum system through physical techniques, followed by condensation and deposition as a thin film on a cooler substrate. PVD is a very versatile method for manufacturing of pure metal films, alloys or compounds of thickness up to 50 μm [67]. At relatively high temperature, a thermal treatment is generally necessary to homogenize the composition of a multilayer deposit [68].

Sputtering and magnetron sputtering (MS) is an evaporation technique used for PVD under vacuum. A sputtering system consists of a vacuum chamber containing a target (a plate of the material to be deposited) and the substrate (i.e., the membrane), in which a sputtering gas (an inert gas such as argon) is introduced to provide the medium in which a glow discharge, or plasma, may be initiated and maintained. Afterwards, positive ions strike the target and remove target atoms and ions by momentum exchange. The condensation of these species over the support produces a thin film. Before sputtering the metal, the ion bombardment of the support is carried out for cleaning its surface and improving the film adherence.

Spray pirolysis is a very simple technique in which a metal salt solution is sprayed into a heated gas stream and, then, pyrolyzed. It could be useful in the case of not requiring very high purity of hydrogen due to the relatively low H_2/other gases perm-selectivity shown by using this technique.

Compared to the other deposition techniques, the spray pyrolysis method shows quite low separation factor, indicating that the technique needs some improvements, especially for producing dense films.

Solvated metal atom deposition method (or co-condensation technique) allows an easy introduction of the metal phase on the inner surface of a tubular membrane. Palladium vapour obtained by the resistive heating of a crucible loaded with Pd shots is co-condensed in a typical glass reactor. At the end of the reaction, the flask is allowed to warm up to −40°C and the resulting yellow-brown solution siphones under argon and handles at low temperature, using the Schlenk tube technique. The amount of palladium in the isolated solution is determined by X-ray fluorescence. Palladium particles are deposited on the inner surface of a tubular membrane by filling the membrane tube (fitted with Teflon stoppers at the ends) with the above solution and heating up to room temperature.

In *Chemical vapour deposition* (CVD) process, a chemical reaction involving a metal complex in the gas phase is performed at a controlled temperature and the produced metal deposits as a thin film by nucleation and growth on the substrate [69]. The deposition takes place on the hot substrate positioned in the CVD reactor. As in the case of PVD technique, the reaction temperature can be reached either by resistive heating of the substrate or by other heating sources [70].

Electrochemical vapour deposition (EVD) is, essentially, a variation of the CVD technique. In the EVD process, for example, a porous substrate separates a mixture of chlorine vapours (ZCl_3, YCl_3, etc.) and an oxygen source (water vapour or oxygen) [11]. Initially, the reactants from both sides of the support inter-diffuse into the pores and form solid oxides, as in the CVD process. When the pores are closed, oxygen ions are conducted across the solid oxide and the oxide film grows on the chlorine side.

In the *Electroplating* (EP) method, a substrate, used as a cathode, is coated with a metal or an alloy in a plating bath [71]. Palladium can be easily deposited in thick and ductile deposits, providing a good control on the composition of the bath, its temperature and current density [72]. The thickness of deposited films can be mastered by controlling electroplating time and current density [73, 74] and film values from a few microns up to millimetres can be obtained. However, the large domains of alloy composition is not easy to control since the relative deposition of two metals simultaneously from the same solution depends on the simplicity of controlling chemical complexing in the bath [11].

The *Electroless plating deposition* (ELP) technique is based upon the controlled auto-catalyzed decomposition or reduction of meta-stable metallic salt complexes on target surfaces [75]. In the case of palladium, usually, the substrate should be pre-seeded with palladium nuclei in an activation solution in order to reduce the induction period of the autocatalytic plating reaction. For some applications, this technique provides strong benefits such as uniformity of deposits on complex shapes and hardness. Palladium and some of its alloys are among the few metals that can be deposited in this way [11]. However, this method presents some drawbacks such as difficult thickness control, costly losses of palladium in the bath, not guaranteed purity of the deposit and so on [75].

Sol–gel technique is a technique adopted for the preparation of thin materials on which the morphological characteristics (e.g., thickness and porosity) must be accurately controlled. Composite membranes resulting from this process are usually microporous and mesoporous, in which permeation of gases is mainly controlled by surface transport and/or the Knudsen flow mechanism [11].

Molecular layering (ML) technique is one of the most promising methods of membrane modification at the atomic level [76, 77]. The ML method is based on the chemisorption of reagents on a solid substrate surface and consists of the irreversible interaction of low-molecular reagents and functional groups of a solid substrate surface under the conditions of continuous reagent feed and the subsequent removal of the formed gaseous products.

However, each aforementioned method presents benefits and drawbacks. For example, both CVD and ELP techniques are able to coat a complex-shaped component with a uniform thickness layer. Unfortunately, non desired compounds and impurities can be formed and incorporated in the palladium layer, reducing the flux of hydrogen through the film. Moreover, by ELP method, it is not easy to control the thickness of the film. On the contrary, an important benefit of the electroless coating is that it is well suited to applications on available commercial tubular membranes. CVD is not an economic process due to the strict conditions required for the process.

In conclusion, Table 2.3 reports the permeation data of different palladium-based composite membranes, produced, principally, by ELP or CVD techniques. Many parameters are reported in the table: membrane type and thickness, temperature and pressure ranges of the permeation experiments, hydrogen flux,

Table 2.3 Permeation data of different palladium-based membranes reported in the literature

Membrane type	T (°C)	Δp (bar)	δ (μm)	J_{H_2} (mol/m²s)	Pe_{H_2} (mol m/m² s Pa)	α_{H_2/N_2}	Preparation method	References
Pd/PSS	520	1.5	10	1.8×10^{-1}	1.2×10^{-11}	–	ELP	[61]
Ti–Ni–Pd	450	3.0	45	$\sim 3.3 \times 10^{-3}$	1.7×10^{-10}	∞	Cold rolling	[61]
Pd/PSS-YSZ	400	–	7–10	2.5×10^{-2}	4.7×10^{-9}	800–900	ELP	[78]
Pd/Al_2O_3	200	0.1	15	2.2×10^{-1}	3.3×10^{-10}	7	ELP	[79]
Pd/glass	350–500	4.0	2	–	3.4×10^{-12}	1140–12900	ELP	[80]
Pd/Al_2O_3	450	–	4.8	–	1.4×10^{-11}	60	ELP	[81]
Pd/Al_2O_3	300	0.3	2–4	$1.0–2.0 \times 10^{-1}$	$1.3–2.7 \times 10^{-11}$	5000	CVD	[82]
Pd/Al_2O_3	528	–	2–3	–	3.5×10^{-12}	<18	ELP	[83]
Pd/Al_2O_3	400	1.0	5	1.6×10^{-1}	7.8×10^{-12}	100–200	ELP	[84]
Pd/$BaZrO_3$	600	–	41	–	–	5.7	CVD	[85]
Pd/MPSS	500	1.0	6	3.0×10^{-1}	1.8×10^{-11}	–	ELP	[86]
Pd/PNS	500	3.6	–	8.3×10^{-2}	–	3.7	MS	[87]
Pd/ZrO_2/PSS	500	1.0	10	8.3×10^{-2}	8.3×10^{-11}	–	ELP	[88]
Pd/αAl_2O_3	370	2.9	1	4.0×10^{-1}	–	3000–8000	ELP	[89]
Pd_{84}–Cu_{16}/ZrO_2–PSS	480	2.5	5	6.0×10^{-1}	2.6×10^{-9}	∞	ELP	[90]
Pd_{90}–Ag_{10}/αAl_2O_3	200–343	0.8–2.5	20	1.4×10^{-1}	2.5×10^{-11}	30–178	ELP	[91]
Pd–Ag/Al_2O_3	–	1.4	10	1.0×10^{-1}	1.0×10^{-11}	1500	ELP	[92]
Pd–Ag/PSS	400–500	1.0	2–3	3.0×10^{-1}	6.0×10^{-12}	–	ELP	[93]
Pd/αAl_2O_3	550	4.0	11	7.0×10^{-2}	–	~1000	ELP	[94]
Pd-Cu/αAl_2O_3	450	3.5	11	8.0×10^{-1}	2.6×10^{-11}	1150	ELP	[95]

hydrogen permeance, ideal separation factor, preparation method of the metallic thin layer and relative bibliography. Generally, it is possible to state that the porous palladium-based membranes show high hydrogen flux and low hydrogen selectivity, while dense palladium-based membranes show low hydrogen flux and high hydrogen selectivity.

2.5 Reaction Processes Using Palladium-based Membranes

The first use of palladium membranes was registered in the 1866, when Graham used a palladium membrane to separate hydrogen from gases mixtures [96]. In 1915, Snelling patented the hydrogen removal through palladium or platinum tubes from a reactor using a granular catalyst for dehydrogenation reactions [53]. In 1964, Gryaznov proposed as a novel application of palladium-based membranes a method for carrying out simultaneously the evolution and the consumption of hydrogen in a dense tubular palladium reactor, where palladium is permeable only to hydrogen and also serves as a catalyst.

The first commercial application of a dense 23 wt% Pd–Ag membrane was in the 1964, when Johnson Matthey used this membrane for purifying a hydrogen rich-stream [97]. Successively, Johnson Matthey also developed a hydrogen generator constituted by a palladium MR fed with a methanol/water mixture. This plant was used in small scale by British Antarctic Survey in 1975 [98, 99].

Moreover, the first pilot-scale composite palladium MR for direct ultra-pure hydrogen production has been realized by the largest gas company in Tokyo (Tokyo Gas Company Ltd.) [100].

Recently, a reformer and membrane modules (RMM) test plant having the capacity of 20 Nm^3/h of hydrogen has been designed and constructed by Tecnimont KT near Chieti, in Italy. This plant is described in Chap. 10 of this book.

Actually, the palladium-based MR represents an alternative solution to conventional systems for pure hydrogen production to be supplied to a PEMFC.

In fact, the palladium-based MR, combining in only one unit the separation phase with the reaction steps, offers the following advantages with respect to a FBR:

- to combine the chemical reaction and the gas separation in only one system reducing the capital costs,
- to enhance the conversion of equilibrium limited reactions. In fact, by the selective removal of one or more products from the reaction side, the thermodynamic equilibrium restrictions can be overcome, due to the so-called "shift effect",
- to achieve higher conversions than FBRs, operating at the same MR conditions, or, on the contrary, the same conversion but operating at milder conditions,
- to improve products yield and selectivity,
- and, especially, to produce a pure hydrogen stream.

Table 2.4 Chemical reactions for producing pure hydrogen by using a palladium-based MR

Kind of reaction	Membrane	Material
Coupling of hydrogenation and dehydrogenation reactions	Dense	Pd
Decomposition of RuO_4 to $RuO_2 + O_2$	Dense	Pd/Ag
Decomposition of ammonia	Dense	Pd
Dehydrogenation of cyclohexane to benzene	Dense	Pd/Ag
Dehydrogenation of ethylbenzene to styrene	Porous	Pd
Dehydrogenation of ethane to ethylene	Dense	Pd/Ag
Dehydrogenation of isopropyl alcohol to acetone	Dense	Pd
Dehydrogenation of water–gas shift reaction	Dense	Pd, Pd/Ag
	Porous	Pd
Dehydrogenation of *n*-heptane to toluene + benzene	Dense	Pd/Rh
Dehydrogenation of butane to butadiene	Dense	Pd
Dehydrogenation of 1,2-cyclohexanediol	Dense	Pd/Cu
Dry reforming of methane	Dense	Pd-alloy
Hydrogenation of ethylene to ethane	Dense	Pd
Hydrogenation of butadiene	Dense	Pd
Hydrogenation of acetylene	Dense	Pd/Ag
Hydrogenation of butenes	Dense	Pd/Sb
Hydrogenation of diene hydrocarbons	Dense	Pd/Ru
Hydrogenation of phenol to cyclohexanone	Dense	Pd; $Pd_{93}Ni_7$; $Pd_{93}Ru_7$; $Pd_{77}Ag_{23}$
Hydrodealkylation of dimethylnapthalenes	Dense	Pd/Ni
Methane conversion into hydrogen and higher hydrocarbons	Porous	Pd-alloy
Octane reforming	Dense	Pd and Pd-alloy
Partial oxidation of methane	Porous	Pd-alloy
Steam reforming of ethanol	Dense	Pd and Pd-alloy
Steam reforming of methane	Dense	Pd-alloy
Steam reforming of methanol	Dense	Pd and Pd-alloy

In order to produce pure hydrogen stream by using a palladium-based MR, many chemical reactions can be used. Some of them are reported in Table 2.4.

In this chapter, a particular attention will be address to the following reactions:

- methane steam reforming,
- dry reforming of methane,
- water gas shift,
- ethanol steam reforming,
- methanol steam reforming,
- bioglycerol steam reforming,
- acetic acid steam reforming.

Moreover, in order to solve the problems related to the environmental pollution, previously mentioned, it will be interesting to investigate the hydrogen production via reforming reactions of bio-fuels such as methanol, glycerol, ethanol, biogas,

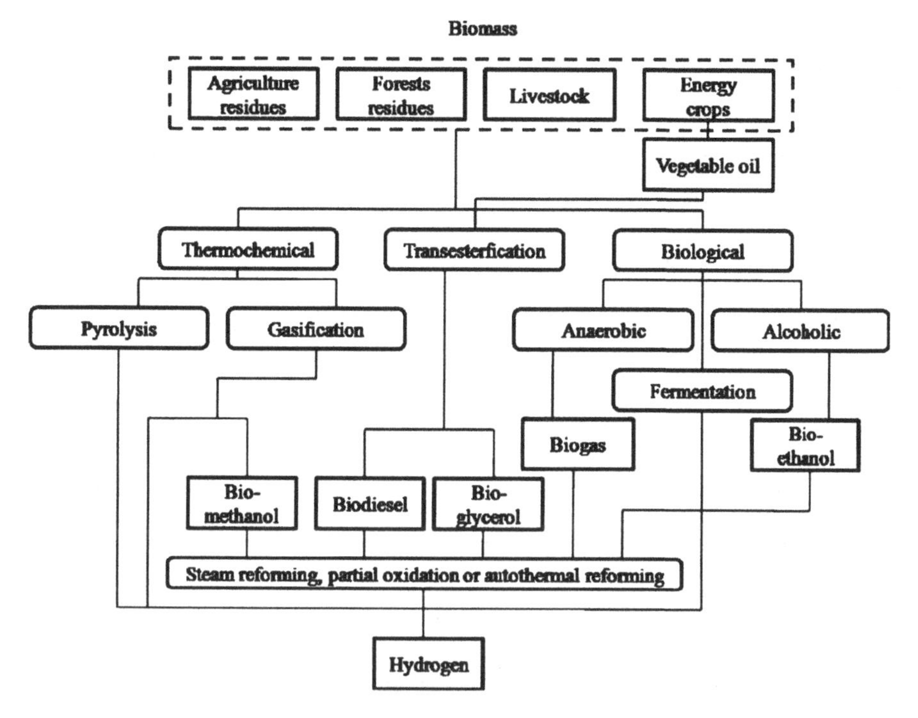

Fig. 2.11 Selected hydrogen production technologies from various renewable sources [101]

etc., which can be produced by renewable sources such as biomass, as reported in Fig. 2.11.

2.5.1 Methane Steam Reforming

Generally, methane steam reforming (MSR) reaction is carried out in FBRs between 800 and 900°C due to the endothermicity of the reaction [27].

$$CH_4 + 2H_2O \leftrightarrows 4H_2 + CO_2 \quad \Delta H^{\circ}_{298\,K} = 165.0\,\text{kJ/mol} \tag{2.9}$$

Only at this elevated temperature, the methane conversion is complete. Furthermore, in the strenuous conditions, the catalyst undergoes deactivation due to carbon formation. As an alternative, using Pd-based MRs it is possible to reach complete methane conversion at lower temperature (~500°C) as summarized in Table 2.5. For example, Lin et al. [102, 103], carried out the MSR reaction in a palladium-based MR, obtaining methane conversions exceeding 80% at temperature between 350 and 500°C rather than 850°C, necessary in a FBR. Chen et al. [104] obtained almost 100% methane conversion at 550°C with respect to 27% obtained in a FBR. In particular, the authors reached 95% pure hydrogen recovery,

Table 2.5 Methane conversion and pure hydrogen recovery data for methane steam reforming reaction

Membrane	T (°C)	$p_{reaction}$ (bar)	MR conversion (%)	FBR conversion (%)	Pure hydrogen recovery (%)	References
Pd–SS	500	20.0	85	–	90	[103]
		9.0	40	20	30	
Pd/Al_2O_3	550	9.0	99	27	95	[104]
Pd/Vycor	500	1.0–9.1	90	–	–	[105]
Pd/5.1%Ag	500	1.4	50	35	–	[106]
Pd–23% Ag	500	6.1	50	~20 (equilibrium)	-	[107]
		10.0	60		-	
Pd-based	500	1.0	100	–	–	[108]
Pd–25%Ag	150	1.0	15.4	12.7	–	[109]
	300		16.5	16.5	–	
Pd–PSS	527	3.0	100	~50	–	[110]
Pd-based	650	2.0–4.0	97	–	–	[111]

confirming that the selective removal of hydrogen from the reaction zone allows to obtain methane conversion significantly higher than a FBR.

Generally, Table 2.5 shows that the palladium-based MRs application allows reducing the operative conditions required for carrying out the MSR reaction in a FBR and to obtain high methane conversion as well as high pure hydrogen recovery.

Chapter 5 will report a detailed assessment of methane steam reforming MR performance. Moreover, it has to be cited the test plant fabricated by Tecnimont KT and described in Chap. 10, which is composed by a series of reformer reactors and Pd-based membrane separators and has a capacity of 20 Nm^3/h of pure H_2 production.

2.5.2 Dry Reforming of Methane

One of the most important drawback related to the methane dry reforming reaction is the carbon deposition.

$$CH_4 + CO_2 = 2CO + 2H_2 \quad \Delta H^\circ_{298\,K} = +247.0\,\mathrm{kJ/mol} \tag{2.10}$$

This was highlighted by different scientists such as Galuszka et al. [112], who observed coke deposition carrying out the reaction (2.10) at 550–600°C in both FBR and MR (housing a dense palladium membrane prepared by electroless plating technique). Gallucci et al. [113] performed the methane dry reforming reaction in FBR and both porous and dense Pd–Ag MRs at 400 and 450°C. The authors demonstrated that the MRs give lower carbon deposition with respect to

Table 2.6 Methane conversion data for methane dry reforming reaction

Membrane	*T* (°C)	MR conversion (%)	FBR conversion (%)	References
Pd-based	550	37.5	17.2	[112]
	600	48.6	40.9	
Porous Pd–Ag	400	2.1	5.6	[113]
	450	8.4	17.4	
Dense Pd–Ag	400	7.9	5.6	[114]
	450	17.8	17.4	
Pd/ceramic composite	500	54	–	[115]

the FBR and, in particular, the lower carbon deposition is obtained when the dense membrane is used.

In this case, the palladium-based MR benefits result in higher methane conversion than FBR, as shown in Table 2.6, and a reduction of the carbon deposition with respect to a conventional reformer, as above reported.

2.5.3 *Water Gas Shift Reaction*

The water gas shift (WGS) reaction is one of the most important industrial processes used to produce hydrogen:

$$CO + H_2O = CO_2 + H_2 \quad \Delta H^{\circ}_{298\,K} = -41.1\,\text{kJ/mol} \tag{2.11}$$

The applications and the research studies performed on this kind of reaction were realized by many scientists. In particular, a great literature is present on this issue concerning the use of palladium-based membrane reactors, as resumed briefly in Table 2.7, where CO conversion values obtained in MR and compared with the thermodynamic equilibrium ones of some scientific works are reported. In particular, among these works, Kikuchi et al. [115] demonstrated that, using a 20 μm layer of palladium-coated onto a porous glass tube produced by the electroless plating method, allows to obtain almost complete CO conversion.

Basile et al. [116] studied the WGS reaction using a MR consisting of a composite palladium-based membrane realized with an ultrathin palladium film (~0.1 μm) coated on the inner surface of a porous ceramic support (γ-Al_2O_3) by the co-condensation technique. The authors pointed out the benefit of applying a palladium MR, taking into account that, at 320°C and 1.1 bar, the thermodynamic equilibrium of CO conversion is around 70%, while the authors obtained with the MR CO conversion of around 100%. Moreover, the same authors illustrated that a complete CO conversion could be reached by using a composite membrane with a thinner palladium layer (10 μm Pd film coated on a ceramic support) [117].

Moreover, Iyoha et al. [118] observed that the CO conversion can decrease from around 90% to around 70% once a palladium MR is replaced with another

Table 2.7 CO conversion data for water gas shift reaction

Membrane	T (°C)	$p_{reaction}$ (bar)	MR conversion (%)	FBR conversion (%)	References
Pd/Vycor	400	1.0	92	76	[115]
Pd/γ-Al_2O_3	320	1.1	100	84	[116]
Pd on ceramic	320	1.0	98	83	[117]
Pure Pd	900	2.4	93	–	[118]
Pd_{80}–Cu_{20}			66		
Pd on porous glass	400	1.0	98	75	[119]
Pd-composite	322	1.1	99.2	99.1	[120]
		1.2	99.9		
Pd–Ag	325	1.0	100	84	[121]
Pure Pd	320	1.3–1.5	100	<50	[122]
Mesoporous Pd			78		
Pd(60%)–Cu	350	–	94	93	[123]

MR containing a $Pd_{80}Cu_{20}$ membrane, due to the lower hydrogen permeance of the Pd/Cu membrane.

In Table 2.7, the comparison between the CO conversion values obtained in different palladium-based MRs and the equilibrium ones is reported, demonstrating the capacity of palladium-based MRs to overcome the thermodynamic limits and to obtain high CO conversion. Chapter 7 reports a detailed assessment of selective membrane application for WGS process.

2.5.4 Ethanol Steam Reforming

Bioethanol is an aqueous solution containing between 8.0 and 12.0 wt% of ethanol and some by-products depending on the raw material used [124]. Nevertheless, the bioethanol distillation is an expensive process, because of the azeotrope presence. For this reason, in the last years, bioethanol is directly used as fuel in steam reforming reaction. Moreover, an excess of water improves the palladium-based MR performances reducing also the CO content as by-products.

$$C_2H_5OH + 3H_2O = 2CO_2 + 6H_2 \quad \Delta H^{\circ}_{298\,K} = +157.0\,kJ/mol \qquad (2.12)$$

A great part of scientific literature is focused on carrying out this reaction using ethanol produced by no-renewable sources [125–130]. Concerning bioethanol steam reforming (BESR) reaction, only few studies are deal with both FBRs [131–133] and MRs use [134–136], as reported in Table 2.8.

In particular, Gernot et al. [134] used a composite Pd-based MR, whose structure consists of a three layers stacking. This membrane structure lowers the thermal expansion stresses between the membrane and the support as well as the quantity of noble metal active layer. At 600°C and for 500 h of work, as best

Table 2.8 Ethanol conversion and pure hydrogen recovery data for bioethanol steam reforming reaction

Membrane	T (°C)	$p_{reaction}$ (bar)	MR conversion (%)	FBR conversion (%)	Pure hydrogen recovery (%)	References
Pd-based	600	–	100	–	–	[134]
Dense Pd–Ag	400	1.5	95	60	30	[135]
Dense Pd–Ag	400	3.0	100	65	95	[136]

results the authors achieved a complete bioethanol conversion as well as a pure hydrogen stream collected with an impurity content <1.0%.

Iulianelli et al. [135, 136] studied from an experimental point of view the steam reforming reaction of a simulated bioethanol mixture (water/ethanol feed molar ratio = 18.7/1 mol/mol without other typical by-products) in a dense Pd–Ag MR in order to produce pure hydrogen. The dense Pd–Ag membrane was prepared by cold rolling and diffusion welding technique [137]. As reported in the table, working at 400°C and 3.0 bar, the authors obtained a complete bioethanol conversion (~85.0% for the FBR working at the same MR operating conditions) and around 95.0% of pure hydrogen recovery.

However, Table 2.8 shows the capacity of palladium-based MRs to obtain higher ethanol conversions than the FBRs.

2.5.5 Methanol Steam Reforming

Methanol is conventionally produced from natural gas [138]. Alternatively, methanol can be obtained from biomass, such as wood and agricultural waste [102]. Renewable methanol presents different advantages as fuel and raw material. For example, it is more easily transportable than methane or other fuel gases, it has high energy density and does not require desulphurization.

Methanol steam reforming (SRM) (3.15) is an endothermic reaction, feasible at temperatures of 200–300°C [31].

$$CH_3OH + H_2O = CO_2 + 3H_2 \quad \Delta H^{\circ}_{298\,K} = +49.7\,kJ/mol \tag{2.13}$$

As indicated in Table 2.9, Wieland et al. [31] carried out the SRM reaction using three different palladium-based MRs (Pd_{75}–Ag_{25}, Pd_{60}–Cu_{40} and Pd–V–Pd). Owing to the low stability of the vanadium-based membrane at pressures above 4.2 bar, the authors compared the performances of the SRM reaction when carried out in a Pd–Ag and a Pd–Cu MRs. The authors showed that a higher pure hydrogen recovery is obtained using the Pd–Ag MR (at 25.0 bar, 300°C, the hydrogen recovery of Pd–Ag MR is 96.0% against 78.0% of the Pd–Cu MR).

Table 2.9 Methanol conversion and pure hydrogen recovery data for methanol steam reforming reaction

Membrane	T (°C)	$p_{reaction}$ (bar)	MR conversion (%)	FBR conversion (%)	Pure hydrogen recovery (%)	References
Pd–Ag	300	25.0	–	–	96	[31]
Pd–Cu					78	
Composite Pd–Ag	450	5.2	76		15	[33]
	550	5.2	73	–	45	
		7.9	72		50	
		11.4	71		53	
	600	11.4	75		60	
Dense Pd–Ag	250	1.3	80	40	–	[140]
TiO_2–Al_2O_3 asymmetric porous commercial membrane with a Pd–Ag deposit	350		30		–	
	600		100		–	
Asymmetric porous ceramic membrane with a Pd–Ag deposit	350	1.3	45	86	–	[142]
	550		65		–	
Pine-hole free Pd–Ag thin wall membrane tube	350		87		–	
	450		100		–	
Dense Pd–Ag	300	1.3	100	55	–	[143]
Dense Pd–Ag	300	3.0	–	–	93	[144, 145]

Lin et al. [102, 139] used a double-jacketed supported palladium MR packed with $Cu/ZnO/Al_2O_3$ catalyst. The authors obtained a pure hydrogen recovery over 70.0%, concluding that this process can constitute an alternative solution of on-board hydrogen generation for electric vehicle fuel cells.

Basile et al. [140] compared the performances in terms of methanol conversion and pure hydrogen production with respect to the ones of a FBR. The authors demonstrated that the MR gives a higher methanol conversion at each operating condition investigated. As best results, a 80.0% methanol conversion is reached working at 250°C and 1.3 bar, as reported in Table 2.9.

Arstad et al. [141] used a self-supported, Pd/(23 wt%) Ag-based MR (with a thickness of 1.6 μm) achieving 100.0% the production of pure hydrogen. Hence, the authors concluded that the low-thickness of the palladium-based membrane can represent a fundamental step for reducing the palladium-cost and making competitive the hydrogen separation technologies by palladium-based membrane.

2.5.6 Bioglycerol Steam Reforming

Bioglycerol is a byproduct of biodiesel production [146], which is usually derived from the transesterification of vegetable oil with methanol or ethanol. During the

Table 2.10 Glycerol conversion and pure hydrogen recovery data for glycerol steam reforming reaction

Membrane	T (°C)	$p_{reaction}$ (bar)	MR conversion (%)	FBR conversion (%)	Pure hydrogen recovery (%)	References
Dense Pd–Ag	400	4.0	94	40	60	[148]
Dense Pd–Ag	400	5.0	60	42	58	[149]

process, the oil is mixed with a metallic base (sodium or potassium hydroxide) and an alcohol (methanol or ethanol). The reaction produces methyl or ethyl ester (biodiesel) and glycerol as a byproduct, which can be used as a renewable source [147]. To the best of our knowledge, at moment, only Iulianelli et al. [148, 149] studied glycerol steam reforming (GSR) reaction in a dense Pd–Ag MR, as reported in Table 2.10.

$$C_3H_8O_3 + 3H_2O = 3CO_2 + 7H_2 \quad \Delta H^\circ_{298\,K} = +346.4\,kJ/mol \tag{2.14}$$

The authors studied the catalyst influence on the reactor performances (glycerol conversion and pure hydrogen recovery), using two commercial catalysts: Co/Al_2O_3 and Ru/Al_2O_3. Using the Co/Al_2O_3 catalyst and at 4.0 bar and 400°C, the authors obtained a glycerol conversion of 94.0% and a pure hydrogen recovery higher than 60.0%. On the contrary, using the Ru/Al_2O_3 catalyst, the authors achieved around 20.0% glycerol conversion and 16.0% pure hydrogen recovery at 5.0 bar. Iulianelli justified these low performances as the main drawback due to the combination of ruthenium with an acid support as Al_2O_3, unfavourable for GSR reaction. Moreover, the authors observed that carbon formation, taking place during the reaction, affects negatively the performances of the Pd–Ag membrane in terms of a lower hydrogen permeated flux and catalyst deactivation.

2.5.7 Acetic Acid Steam Reforming

Acetic acid is a renewable source and can be easily obtained by fermentation of biomass [150]. Few studies [151–154] concerned the acetic acid steam reforming (AASR) reaction for producing hydrogen by means only FBRs.

$$CH_3COOH + 2H_2O = 2CO_2 + 4H_2 \quad \Delta H^\circ_{298\,K} = +134.9\,kJ/mol \tag{2.15}$$

Only two scientific papers were published dealing with the use of MR for carrying out the AASR reaction [155, 156], as shown in Table 2.11. In these studies, the AASR reaction was performed in a dense Pd–Ag MR packed with two kinds of catalyst: a Ni-based commercial catalyst in the first case and both

Table 2.11 Acid acetic conversion and pure hydrogen recovery data for acetic acid steam reforming reaction

Membrane	T (°C)	$p_{reaction}$ (bar)	MR conversion (%)	FBR conversion (%)	Pure hydrogen recovery (%)	References
Dense Pd–Ag (MR1)	400 450	2.5	100 100	85 75	32 36	[155]
Dense Pd–Ag (MR2)	400 450		100 100	92 87	26 32	
Dense Pd–Ag	400	4.0	100	–	70	[156]

Ru-based and a Ni-based commercial catalyst in the second case. In both experiments, a complete acetic acid conversion was obtained.

2.6 Conclusions

An extensive overview concerning palladium-based membranes was presented, subsequently to a general classification of the membranes. In particular, an assessment of the problems associated with the pure palladium membranes was presented, which was followed by a description of the preparation methods of palladium-based membranes and their industrial applications.

In particular, this chapter highlighted the importance of palladium-based membranes for producing pure hydrogen. Applicability of these membranes is limited by sensitivity towards certain species and cost. When these membranes are applied to the reactor system, the MRs constitute an interesting alternative approach to the FBRs, owing to the ability of the palladium-based MRs in performing simultaneously the reaction process and the selective hydrogen separation. In this route, the continuous hydrogen removal permits to obtain higher reaction conversions than the thermodynamic equilibrium, which is the upper limit to be considered in a FBR. In the meanwhile, a pure hydrogen stream for directly feeding a PEMFC without needing any other purification process can be obtained.

References

1. http://scopees.elsevier.com
2. Juenker DW, Van Swaay M, Birchenall CE (1955) On the use of palladium diffusion membranes for the purification of hydrogen. Rev Sci Instrum 26:888
3. Gao H, Lin YS, Li Y, Zhang B (2004) Chemical stability and its improvement of palladium-based metallic membranes. Ind Eng Chem Res 43:6920–6930
4. Goltsov V, Veziroglu N (2001) From hydrogen economy to hydrogen civilization. Int J Hydrogen Energy 26:909–915

5. Koros WJ, Ma YH, Shimidzu T (1996) Terminology for membranes and membrane processes. J Memb Sci 120:149–159
6. Khulbe KC, Feng CY, Matsuura T (2007) Synthetic polymeric membranes, characterization by atomic force microscopy. Springer, pp 216, ISBN:3540739939
7. Xia Y, Lu Y, Kamata K, Gates B, Yin Y (2003) Macroporous materials containing three-dimensionally periodic structures. In: Yang P (ed) Chemistry of nanostructured materials. World Scientific, Singapore, pp 69–100
8. Catalytica® (1988) Catalytic membrane reactors: concepts and applications, Catalytica Study N. 4187 MR
9. Van Veen HM, Bracht M, Hamoen E, Alderliesten PT (1996) Feasibility of the application of porous inorganic gas separation membranes in some large-scale chemical processes. In: Burggraaf AJ, Cot L (eds) Fundamentals of inorganic membrane science and technology, vol 14. Elsevier, New York, pp 641–681
10. Mallevialle J, Odendaal PE, Wiesner MR (eds) (1998) Water treatment membrane processes. McGraw-Hill Publishers, New York
11. Mulder M (1996) Basic principles of membrane technology. Kluwer Academic, Dordrecht, p 564
12. Saracco G, Specchia V (1994) Catalytic inorganic membrane reactors: present experience and future opportunities. Catal Rev Sci Eng 36:305–384
13. Knozinger H, Ratnasamy P (1978) Catalytic aluminas: surface models and characterization of surface sites. Catal Rev Sci Eng 17:31–70
14. Kapoor A, Yang RT, Wong C (1989) Surface diffusion. Catal Rev 31:129–214
15. Falconer JL, Noble RD, Sperry DP (1995) Catalytic membrane reactors. In: Noble RD, Stern SA (eds) Membrane separations technology: principles and applications. Elsevier, New York, pp 669–712
16. Sperry DP, Falconer JL, Noble RD (1991) Methanol–hydrogen separation by capillary condensation in inorganic membranes. J Memb Sci 60:185–193
17. Lee KH, Hwang ST (1986) Transport of condensible vapors through a microporous vycor glass membrane. J Coll Int Sci 110:544–555
18. Ulhorn RJR, Keizer K, Burggraaf AJ (1992) Gas transport and separation with ceramic membranes. Part I. Multilayer diffusion and capillary condensation. J Memb Sci 66:259–269
19. Adhikari S, Fernand S (2006) Hydrogen membrane separation techniques. Ind Eng Chem Res 45:875–881
20. Stambouli A, Traversa E (2002) Fuel cells, an alternative to standard sources of energy. Renew Sust Energy Rev 6:295–304
21. Mehta V, Cooper JS (2003) Review and analysis of PEM fuel cell design and manufacturing. J Power Sources 114:32–53
22. Costamagna P, Srinivasan S (2001) Quantum jumps in the PEMFC science and technology from the 1960s to the year 2000. Part II. Engineering, technology, development and application aspects. J Power Sources 102:253–269
23. http://ec.europa.eu/research/rtdinfo/42/01/article_1315_en.html Accessed 24 Nov 2010
24. http://www.hydrogenassociation.org/general Accessed 24 Nov 2010
25. Rifkin J (2002) The hydrogen economy: the creation of the worldwide energy web and the redistribution of power on earth. Jeremy P. Tarcher, Penguin, ISBN 1-58542-193-6
26. Cheng X, Shi Z, Glass N, Zhang L, Zhang J, Song D, Liu ZS, Wang H, Shen J (2007) A review of PEM hydrogen fuel cell contamination: impacts, mechanisms, and mitigation. J Power Sources 165:739–756
27. Barelli L, Bidini G, Gallorini F, Servili S (2008) Hydrogen production through sorption-enhanced steam methane reforming and membrane technology: a review. Energy 33:554–570
28. Basile A (2008) Hydrogen production using Pd-based membrane reactors for fuel cells. Top Catal 51:107–122
29. Tosti S, Bettinali L, Violante V (2000) Rolled thin Pd and Pd–Ag membranes for hydrogen separation and production. Int J Hydrogen Energy 25:319–325

30. Cheng YS, Pena MA, Fierro JL, Hui DCW, Yeung KL (2002) Performance of alumina, zeolite, palladium, Pd–Ag alloy membranes for hydrogen separation from Towngas mixture. J Memb Sci 204:329–340
31. Wieland S, Melin T, Lamm A (2002) Membrane reactors for hydrogen production. Chem Eng Sci 57:1571–1576
32. Valenti G, Macchi F (2008) Proposal of an innovative, high efficiency, large-scale hydrogen liquefier. Int J Hydrogen Energy 33:3116–3121
33. Damle AS (2009) Hydrogen production by reforming of liquid hydrocarbons in a membrane reactor for portable power generation—experimental studies. J Power Sources 186:167–177
34. Gallucci F, De Falco M, Tosti S, Marrelli L, Basile A (2007) The effect of the hydrogen flux pressure and temperature dependence factors on the membrane reactor performances. Int J Hydrogen Energy 32:4052–4058
35. Koros WJ, Fleming GK (1993) Membrane-based gas separation. J Memb Sci 83:1–80
36. Dolan MD, Dave NC, Ilyushechkin AY, Morpeth LD, McLennan KG (2006) Composition and operation of hydrogen-selective amorphous alloy membranes. J Memb Sci 285:30–55
37. Gallucci F, Tosti S, Basile A (2008) Synthesis, characterization and applications of palladium membranes. In: Mallada R, Menendez M (eds) Inorganic membranes: synthesis, characterization and applications, Chapter 8. Elsevier, Amsterdam (Netherlands)
38. Grashoff GJ, Pilkington CE, Corti CW (1983) The purification of hydrogen—a review of the technology emphasing, the current status of palladium membrane diffusion. Platinum Met Rev 27:157–168
39. Lewis FA, Kandasamy K, Baranowski B (1988) The "Uphill" diffusion of hydrogen—strain-gradient-induced effects in palladium alloy membranes. Platinum Met Rev 32:22–26
40. Hsieh HP (1989) Inorganic membrane reactors—a review. AIChE Symp Ser 85:53–67
41. Brodowsky H (1972) On the non-ideal solution behavior of hydrogen in metals. Ber Bunsenges Physik Chem 76:740–749
42. Shu J, Grandjean BPA, Van Neste A, Kaliaguine S (1991) Catalytic palladium-based membrane reactors: a review. Can J Chem Eng 69:1036–1060
43. Edlund DJ, Pledger WA (1993) Thermolysis of hydrogen sulfide in a metal-membrane reactor. J Memb Sci 77:255–264
44. Edlund DJ, Pledger WA (1994) Catalytic platinum-based membrane reactor for removal of H_2S from natural gas streams. J Memb Sci 94:111–119
45. Noordermeer A, Kok GA, Nieuwenhuys BE (1986) Comparison between the adsorption properties of Pd (111) and PdCu (111) surfaces for carbon monoxide and hydrogen. Surf Sci 172:349–362
46. Li A, Liang W, Hughes R (2000) The effect of carbon monoxide and steam on the hydrogen permeability of a Pd/stainless steel membrane. J Memb Sci 165:135–141
47. Amandusson H, Ekedahl LG, Dannetun H (2000) The effect of CO and O_2 on hydrogen permeation through a palladium membrane. Appl Surf Sci 153:259–267
48. Heras JM, Estiù G, Viscido L (1997) The interaction of water with clean palladium films: thermal desorption and work fucnction study. Appl Surf Sci 108:455–464
49. McCool BA, Lin YS (2001) Nanostructured thin palladium-silver membranes: effects of grain size on gas permeation properties. J Mater Sci 36:3221–3227
50. Nishimura C, Komaki M, Hwang S, Amano M (2002) V–Ni alloy membranes for hydrogen purification. J Alloys Compd 330–332:902–906
51. Luo W, Ishikawa K, Aoki K (2006) High hydrogen permeability in the Nb-rich Nb–Ti–Ni alloy. J Alloys Compd 407:115–117
52. Adams TM, Mickalonis J (2007) Hydrogen permeability of multiphase V–Ti–Ni metallic membranes. Mater Lett 61:817–820
53. Mallada R, Menéndez M (eds) (2008) Inorganic membranes: synthesis, characterization and applications, Elsevier, Amsterdam (Netherlands)
54. Uemiya S (1999) State-of-art of supported metal membranes for gas separation. Sep Purity Methods 28:51–85

55. Tosti S, Borelli R, Borgognoni F, Favuzza P, Rizzello C, Tarquini P (2008) Study of a dense metal membrane reactor for hydrogen separation from hydroiodic acid decomposition. Int J Hydrogen Energy 33:5106–5114
56. Howard BH, Killmeyer RP, Rothenberger KS, Cugini AV, Morreale BD, Enick RM, Bustamante F (2004) Hydrogen permeance of palladium–copper alloy membranes over a wide range of temperatures and pressures. J Memb Sci 241:207–218
57. Gryaznov VM (2000) Metal containing membranes for the production of ultrapure hydrogen and the recovery of hydrogen isotopes. Sep Purity Methods 29:171–187
58. Hwang ST, Kammermeyer K (1975) Techniques in chemistry: membranes in separation. Wiley Interscience, New York
59. McKinley DL, Nitro WV (1967) Metal alloy for hydrogen separation and purification. US patent 3,350,845
60. Hara S, Hatakeyama N, Itoh N, Kimura HM, Inoue A (2002) Hydrogen permeation through palladium-coated amorphous Zr–M–Ni (M = Ti, Hf) alloy membranes. Desalination 144:115–120
61. Basile A, Gallucci F, Iulianelli A, Tereschenko GF, Ermilova MM, Orekhova NV (2008) Ti–Ni–Pd dense membranes—the effect of the gas mixtures on the hydrogen permeation. J Memb Sci 310:44–50
62. Criscuoli A, Basile A, Drioli E, Loiacono O (2001) An economic feasibility study for water gas shift membrane reactor. J Memb Sci 181:21–27
63. Wilde G, Dinda GP, Rösner H (2005) Synthesis of bulk nanocrystalline materials by repeated cold rolling. Adv Eng Mat 7:11–15
64. Smith GV, Brower WE, Matyjaszczyk MS, Pettit TL (1981) Metallic glasses: new catalyst systems. In: Seiyama T, Tanabe K (eds) Proceedings of the 7th international congress on catalysis part A, Elsevier, New York, pp 355–363
65. Molnar A, Smith GV, Bartok M (1989) New catalytic materials from amorphous metal alloys. Adv Catal 36:329–383
66. Kishimoto S, Yoshida N, Arita Y, Flanagan TB (1990) Solution of hydrogen in cold-worked and annuale $Pd_{95}Ag_5$ alloys. Ber Bunsenges Physik Chem 94:612–615
67. Reichelt K, Jiang X (1990) The preparation of thin films by physical vapor deposition methods. Thin Solid Films 191:91–126
68. Mattox DM (1998) Handbook of physical vapor deposition (PVD) processing: film formation, adhesion, surface preparation and contamination control. Noyes Publications, Westwood, NJ, ISBN 0815514220
69. Jones AC, Hitchman ML (2008) Chemical vapour deposition precursors, processes and applications. RSC Publishing, London. doi:10.1039/9781847558794
70. Biswas DR (1986) Review: deposition processes for films and coatings. J Mater Sci 21:2217–2223
71. Mohler JB (1969) Electroplating and related processes. Chemical Publishing Co., New York, ISBN 0-8206-0037-7
72. Wise EM (1968) Palladium-recovery properties and uses. Academic Press, New York
73. Sturzenegger B, Puippe JC (1984) Electrodeposition of palladium–silver alloys from ammoniacal electrolytes. Platinum Met Rev 20:117–124
74. Reid HR (1985) Palladium–nickel electroplating. Effects of solution parameters on alloy properties. Platinum Met Rev 29:61–62
75. Loweheim FA (1974) Modern electroplating. Wiley, New York, pp 342–357 and 739–747
76. Malygin AA (2006) The molecular layering nanotechnology: basis and application. J Ind Eng Chem 12:1–11
77. Tereshchenko GF, Orekhova NV, Ermilova MM, Malygin AA, Orlova AI (2006) Nanostructured phosphorus–oxide-containing composite membrane catalysts. Catal Today 118:85–89
78. Huang Y, Dittmeyer R (2007) Preparation of thin palladium membranes on a porous support with rough surface. J Memb Sci 302:160–170

79. Altinisik O, Dogan M, Dogu G (2005) Preparation and characterization of palladium-plated porous glass for hydrogen enrichment. Catal Today 105:641–646
80. Wang D, Flanagan TB, Shanahan KL (2004) Permeation of hydrogen through pre-oxidized membranes in presence and absence of CO. J Alloys Compd 372:158–164
81. Van Dyk L, Miachon S, Lorenzen L, Torres M, Fiaty K, Dalmon JA (2003) Comparison of microporous MFI and dense Pd membrane performances in an extractor-type CMR. Catal Today 82:167–177
82. Itoh N, Akiha T, Sato T (2005) Preparation of thin palladium composite membrane tube by a CVD technique and its hydrogen permselectivity. Catal Today 104:231–237
83. Kleinert A, Grubert G, Pan X, Hamel C, Seidel-Morgenstern A, Caro J (2005) Compatibility of hydrogen transfer via Pd-membranes with the rates of heterogeneously catalysed steam reforming. Catal Today 104:267–273
84. Liang W, Hughes R (2005) The effect of diffusion direction on the permeation rate of hydrogen in palladium composite membranes. Chem Eng J 112:81–86
85. Okada S, Mineshige A, Kikuchi T, Kobune M, Yazawa T (2007) Cermet-type hydrogen separation membrane obtained from fine particles of high temperature proton-conductive oxide and palladium. Thin Solid Films 515:7342–7346
86. Tong J, Suda H, Haraya K, Matsumura Y (2005) A novel method for the preparation of thin dense Pd membrane on macroporous stainless steel tube filter. J Memb Sci 260:10–18
87. Ryi SK, Park JS, Kim SH, Cho SH, Park JS, Kim DW (2006) Development of a new porous metal support of metallic dense membrane for hydrogen separation. J Memb Sci 279:439–445
88. Wang D, Tong J, Xu H, Matsamura Y (2004) Preparation of palladium membrane over porous stainless steel tube modified with zirconium oxide. Catal Today 93–95:689–693
89. Nair BKR, Harold MP (2007) Pd encapsulated and nanopore hollow fiber membranes: synthesis and permeation studies. J Memb Sci 290:182–195
90. Gao H, Lin JYS, Li Y, Zhang B (2005) Electroless plating synthesis, characterization and permeation properties of Pd–Cu membranes supported on ZrO_2 modified porous stainless steel. J Memb Sci 265:142–152
91. Huang TC, Wei MC, Chen HI (2003) Preparation of hydrogen-permselective palladium–silver alloy composite membranes by electroless co-deposition. Sep Purif Technol 32:239–245
92. Liang W, Hughes R (2005) The catalytic dehydrogenation of isobutane to isobutene in a palladium/silver composite membrane reactor. Catal Today 104:238–243
93. Tong J, Shirai R, Kashima Y, Matsumura Y (2005) Preparation of a pinhole-free Pd–Ag membrane on a porous metal support for pure hydrogen separation. J Memb Sci 260:84–89
94. Nair BKR, Choi J, Harold MP (2007) Electroless plating and permeation features of Pd and Pd/Ag hollow fiber composite membranes. J Memb Sci 288:67–84
95. Roa F, Douglas WJ, Mc Cormik RL, Paglieri SN (2003) Preparation and characterization of Pd–Cu composite membranes for hydrogen separation. Chem Eng J 93:11–22
96. Graham T (1866) On the absorption and dialytic separation of gases by colloid septa, Phil R Soc Lond 156:399–439
97. Booth JCS, Doyle VL, Gee SM, Miller J, Scholtz LA, Walker PA (1996) Advanced hydrogen separation via thin Pd membranes. In: Proceedings of the 11th World Hydrogen Energy Conference, Stuttgart, Germany, pp 867–878
98. Cole MJ (1981) The generator of pure hydrogen for industrial applications. Platinum Met Rev 25:12–13
99. Philpott J (1985) Hydrogen diffusion technology. Commercial applications of palladium membrane. Platinum Met Rev 29:12–16
100. Paturzo L, Basile A, Drioli E (2002) High temperature membrane reactors and integrated membrane operations. Rev Chem Eng 18:511–551
101. Xuan J, Leung MKH, Leung DYC, Ni M (2009) A review of biomass-derived fuel processors for fuel cell systems. Renew Sust Energy Rev 13:1301–1313
102. Lin YM, Rei MH (2000) Process development for generating high purity hydrogen by using supported membrane reactor as steam reformer. Int J Hydrogen Energy 25:211–219

103. Lin YM, Liu SL, Chuang CH, Chu YT (2003) Effect of incipient removal of hydrogen through palladium membrane on the conversion of methane steam reforming. Experimental and modeling. Catal Today 82:127–139
104. Chen Y, Wang Y, Xu H, Xiong G (2008) Efficient production of hydrogen from natural gas steam reforming in palladium membrane reactor. Appl Catal B 80:283–294
105. Uemiya S, Sato N, Ando H, Matsuda T, Kikuchi E (1991) Steam reforming of methane in a hydrogen-permeable membrane reactor. Appl Catal 67:223–230
106. Shu J, Grandjean BPA, Kaliaguine S (1995) Asymmetric Pd–Ag/stainless steel catalytic membranes for methane steam reforming. Catal Today 25:327–332
107. Jorgensen S, Nielsen PEH, Lehrmann P (1995) Steam reforming of methane in membrane reactor. Catal Today 25:303–307
108. Kikuchi E, Nemoto Y, Kajiwara M, Uemiya S, Kojima T (2000) Steam reforming of methane in membrane reactors: comparison of electroless-plating and CVD membranes and catalyst packing modes. Catal Today 56:75–81
109. Basile A, Paturzo L, Vazzana A (2003) Membrane reactor for the production of hydrogen and higher hydrocarbons from methane over Ru/Al_2O_3 catalyst. Chem Eng J 93:31–39
110. Tong J, Matsumura Y (2005) Effect of catalytic activity on methane steam reforming in hydrogen-permeable membrane reactor. Appl Catal A 286:226–231
111. Patil CS, Annaland MVS, Kuipers JAM (2007) Fluidised bed membrane reactor for ultrapure hydrogen production via methane steam reforming: experimental demonstration and model validation. Chem Eng Sci 62:2989–3007
112. Galuszka J, Pandey RN, Ahmed S (1998) Methane conversion to syngas in a palladium membrane reactor. Catal Today 46:83–89
113. Gallucci F, Tosti S, Basile A (2008) Pd–Ag tubular membrane reactors for methane dry reforming: a reactive method for CO_2 consumption and H_2 production. J Memb Sci 317:96–105
114. Kikuchi E (1995) Palladium/ceramic membranes for selective hydrogen permeation and their application to membrane reactor. Catal Today 25:333–337
115. Kikuchi E, Uemiya S, Sato N, Inoue H, Ando H, Matsuda T (1989) Membrane reactor using microporous glass supported thin film of palladium. Application to the water gas shift reaction. Chem Lett 18:489–492
116. Basile A, Chiappetta G, Tosti S, Violante V (2001) Experimental and simulation of both Pd and Pd/Ag for a water gas shift membrane reactor. Sep Purif Technol 25:549–571
117. Basile A, Violante V, Santella F, Drioli E (1995) Membrane integrated system in the fusion reactor fuel cycle. Catal Today 25(3–4):321–326
118. Iyoha O, Enick R, Killmeyer R, Howard B, Morreale B, Ciocco M (2007) Wall-catalyzed water-gas shift reaction in multi-tubular Pd and 80 wt%Pd–20 wt%Cu membrane reactors at 1173 K. J Memb Sci 298:14–23
119. Uemiya S, Sato N, Ando H, Kikuchi E (1991) The water gas shift reaction assisted by a palladium membrane reactor. Ind Eng Chem Res 30:585–589
120. Basile A, Criscuoli A, Santella F, Drioli E (1996) Membrane reactor for water gas shift reaction. Gas Sep Purif 10:243–254
121. Tosti S, Violante V, Basile A, Chiappetta G, Castelli S, De Francesco M, Scaglione S, Sarto F (2000) Catalytic membrane reactors for tritium recovery from tritiated water in the ITER fuel cycle. Fusion Eng Des 49–50:953–958
122. Criscuoli A, Basile A, Drioli E (2000) An analysis of the performance of membrane reactors for the water–gas shift reaction using gas feed mixtures. Catal Today 56:53–64
123. Flytzani-Stephanopoulos M, Qi X, Kronewitter S (2004) Water–gas shift with integrated hydrogen separation process. Final report to DOE, Grant # DEFG2600-NT40819, pp 1–38
124. Pfeffer M, Wukovits W, Beckmann G, Friedl A (2007) Analysis and decrease of the energy demand of bioethanol-production by process integration. Appl Thermal Eng 27:2657–2664
125. Haga F, Nakajima T, Yamashita K, Mishima S (1998) Effect of cristallite size on the catalysis of alumina-supported cobalt catalyst for steam reforming of ethanol. React Kinet Catal Lett 63:253–259

126. Llorca J, Homs N, Sales J, de la Piscina PR (2002) Efficient production of hydrogen over supported cobalt catalysts from ethanol steam reforming. J Catal 209:306–317
127. Batista MS, Santos RKS, Assaf EM, Assaf JM, Ticianelli EA (2003) Characterization of the activity and stability of supported cobalt catalysts for the steam reforming of ethanol. J Power Sources 124:99–103
128. Haryanto A, Fernando S, Murali N, Adhikari S (2005) Current status of hydrogen production techniques by steam reforming of ethanol: a review. Energy Fuels 19:2098–2106
129. Basile A, Gallucci F, Iulianelli A, Tosti S, Drioli E (2006) The pressure effect on ethanol steam reforming in membrane reactor: experimental study. Desalination 200:671–672
130. Basile A, Gallucci F, Iulianelli A, Tosti S (2008) CO-free hydrogen production by ethanol steam reforming in a Pd–Ag membrane reactor. Fuel Cells 1:62–68
131. Benito M, Sanz JL, Isabel R, Padilla R, Arjona R, Daza L (2005) Bio-ethanol steam reforming: Insights on the mechanism for hydrogen production. J Power Sources 151:11–17
132. Dolgykh L, Stolyarchuk I, Denyega I, Strizhak P (2006) The use of industrial dehydrogenation catalyst for hydrogen production from bioethanol. Int J Hydrogen Energy 31:1607–1610
133. Frusteri F, Freni S, Chiodo V, Donato S, Bonura G, Cavallaro S (2006) Steam and auto-thermal reforming of bio-ethanol over MgO and CeO_2 Ni supported catalysts. Int J Hydrogen Energy 31:2193–2199
134. Gernot E, Aupretre F, Deschamps A, Epron F, Merecot P, Duprez D, Etievant C (2006) Production of hydrogen from bioethanol in catalytic membrane reactor. 16th Confèrence Mondiale de l'Hydrogène Energie (WHEC16), Lyon (France) http://www.ceth.fr/download/presse/art_ceth_3.pdf
135. Iulianelli A, Liguori S, Longo T, Tosti S, Pinacci P, Basile A (2010) An experimental study on bio-ethanol steam reforming in a catalytic membrane reactor. Part II: reaction pressure, sweep factor and WHSV effects. Int J Hydrogen Energy. 35:3159–3164
136. Iulianelli A, Basile A (2010) An experimental study on bio-ethanol steam reforming in a catalytic membrane reactor. Part I: temperature and sweep-gas flow configuration effects. Int J Hydrogen Energy. 35:3170–3177
137. Tosti S, Bettinali L (2004) Diffusion bonding of Pd–Ag rolled membranes. J Mater Sci 39:3041–3046
138. Cifre GP, Badr O (2007) Renewable hydrogen utilization for the production of methanol. Energy Conver Manag 48:519–527
139. Lin YM, Rei MH (2001) Study on the hydrogen production from methanol steam reforming in supported palladium membrane reactor. Catal Today 67:77–84
140. Basile A, Gallucci F, Paturzo L (2005) A dense Pd/Ag membrane reactor for methanol steam reforming: experimental study. Catal Today 104:244–250
141. Arstad B, Venvik H, Klette H, Walmsley JC, Tucho WM, Holmestad R, Holmen A, Bredesen R (2006) Studies of self-supported 1.6 μm Pd/23 wt% Ag membranes during and after hydrogen production in a catalytic membrane reactor. Catal Today 118:63–72
142. Basile A, Tosti S, Capannelli G, Vitulli G, Iulianelli A, Gallucci F, Drioli E (2006) Co-current and counter-current modes for methanol steam reforming membrane reactor: experimental study. Catal Today 118:237–245
143. Basile A, Parmaliana A, Tosti S, Iulianelli A, Gallucci F, Espro C, Spooren J (2008) Hydrogen production by methanol steam reforming carried out in membrane reactor on Cu/Zn/Mg-based catalyst. Catal Today 137:17–22
144. Iulianelli A, Longo T, Basile A (2008) Methanol steam reforming in a dense Pd–Ag membrane reactor: the pressure and WHSV effects on CO-free H_2 production. J Memb Sci 323:235–240
145. Iulianelli A, Longo T, Basile A (2008) Methanol steam reforming reaction in a Pd–Ag membrane reactor for CO-free hydrogen production. Int J Hydrogen Energy 33:5583–5588
146. Valliyappan T, Ferdous D, Bakhshi NN, Dalai AK (2008) Production of hydrogen and syngas via steam gasification of glycerol in a fixed-bed reactor. Top Catal 49:59–67

147. Adams J, Cassarino C, Lindstrom J, Spangler L, Binder MJ, Holcomb FH (2004) Canola oil fuel cell demonstration I. US Army Corps of Engineers, Washington, DC
148. Iulianelli A, Longo T, Liguori S, Basile A (2010) Production of hydrogen via glycerol steam reforming in a Pd–Ag membrane reactor over Co–Al_2O_3 catalyst. Asia Pac J Chem Eng. doi:10.1002/apj.365
149. Iulianelli A, Seelam PK, Liguori S, Longo T, Keiski R, Calabrò V, Basile A (2010) Hydrogen production for PEM fuel cell by gas phase reforming of glycerol as byproduct of bio-diesel. The use of a Pd–Ag membrane reactor at middle reaction temperature. Int J Hydrogen Energy
150. Liu BF, Ren NQ, Tang J, Ding J, Liu WZ, Xu JF, Cao GL, Guo WQ, Xie GJ (2009) Bio-hydrogen production by mixed culture of photo- and dark-fermentation bacteria. Int J Hydrogen Energy. doi:10.1016/j.ijhydene.2009.05.005
151. Takanabe K, Aika K, Seshanb K, Lefferts L (2004) Sustainable hydrogen from bio-oil-steam reforming of acetic acid as a model oxygenate. J Catal 227:101–108
152. Hu X, Lu G (2007) Investigation of steam reforming of acetic acid to hydrogen over Ni–Co metal catalyst. J Mol Catal A 261:43–48
153. Bimbela F, Oliva M, Ruiz J, Garc'ıa L, Arauzo J (2007) Hydrogen production by catalytic steam reforming of acetic acid, a model compound of biomass pyrolysis liquids. J Anal Appl Pyrolysis 79:112–120
154. Basagiannis AC, Verykios XE (2007) Catalytic steam reforming of acetic acid for hydrogen production. Int J Hydrogen Energy 32:3343–3355
155. Basile A, Gallucci F, Iulianelli A, Borgognoni F, Tosti S (2008) Acetic acid steam reforming in a Pd–Ag membrane reactor: the effect of the catalytic bed pattern. J Memb Sci 311:46–52
156. Iulianelli A, Longo T, Basile A (2008) CO-free hydrogen production by steam reforming of acetic acid carried out in a Pd–Ag membrane reactor: the effect of co-current and counter-current mode. Int J Hydrogen Energy 33:4091–4096

Chapter 3
Hydrogen Palladium Selective Membranes: An Economic Perspective

G. Iaquaniello, A. Borruto, E. Lollobattista, G. Narducci and D. Katsir

3.1 Introduction

In the past decade, large demand in the production of hydrogen for petroleum and petrochemical industry and the potential future demand for hydrogen economy have created considerable interest and efforts to develop more efficient processes for hydrogen separation and production. H_2 separation and production by membranes and membrane reactors at high temperatures and pressures was and still seems to be one of the most promising developing areas. Although the major development effort, if dense Pd membranes are excluded due to their low permeation flux, thin palladium films supported on porous substrates have not yet reached a commercial stage due to some technical issues as long term permeance and selectivity stability, but also to the cost related to their manufacture. In this chapter, a short review of palladium membranes characteristics, hydrogen transport phenomena together with deposition techniques is provided.

Moreover, a review of membranes and "membrane modules" actually available on the market, not comprehensive of all the suppliers, allows focusing on the main costs determinants in the membranes production cycle.

G. Iaquaniello (✉)
Tecnimont-KT S.p.A, Viale Castello della Magliana 75, 00148 Rome, Italy
e-mail: Iaquaniello.G@tecnimontkt.it

A. Borruto and G. Narducci
Department of Chemical Engineering, Materials and Environment, University of Rome La Sapienza, via Eudossiana 18, 00184 Rome, Italy
e-mail: adelina.borruto@uniroma1.it

E. Lollobattista
Processi Innovativi S.r.l., Corso Federico 36, 67100 L'Aquila, Italy
e-mail: lollobattista.e@processiinnovativi.it

D. Katsir
Acktar Ltd., 1 Leshem St, P.O.B. 8643, 82000 Kiryat-Gat, Israel
e-mail: dina.katsir@acktar.com

M. De Falco et al. (eds.), *Membrane Reactors for Hydrogen Production Processes*,
DOI: 10.1007/978-0-85729-151-6_3,

On such a basis, a comparison of film and supports deposition technologies was illustrated to better understand how from a lab technique it is possible to move to an industrial manufacturing process and reach a volume high enough to sustain important economics of scale.

3.2 Membranes Characteristics

A suitable hydrogen membrane has to have the following features: high selectivity for hydrogen, high permeability in order to operate with high flows and limited surfaces, good chemical and structural stability in order to avoid deterioration under exertion.

The more interesting configurations for a hydrogen selective membrane are then:

1. ceramic support + thin hydrogen selective layer
2. metallic support + metallic interdiffusion barrier + thin hydrogen selective layer

Two different geometries, planar and tubular (Fig. 3.1), can be realized from a practical point of view:

3.2.1 Membrane Support

The support gives the membranes the necessary mechanical strength and it usually does not show selective properties.

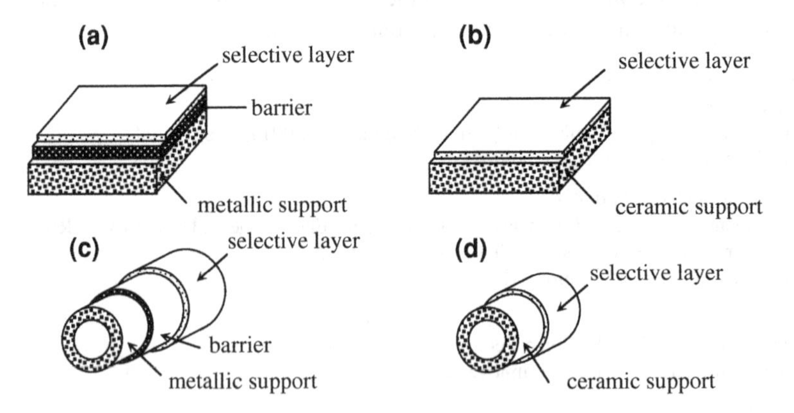

Fig. 3.1 Composite membranes: **a–b** planar configuration with or without barrier, **c–d** tubular configuration with or without barrier

Clearly a support, as part of the membrane, has to be gases permeable and has to be characterized from a specific interconnecting porosity to allow the gases crossing. Therefore, a suitable support has to be realized with a thickness of a few millimeters of sintered materials, either metallic or ceramic ones [1]:

- Metallic materials (as sintered stainless steel) are advantageous due to their thermal expansion coefficient close to that of Pd-based films, removing the disbonding phenomenon due to differential thermal expansion. Moreover, this materials show a relevant weldability and a low cost [1, 2].
  Among the different steels used for these applications, we report as an example: AISI316L [3, 4] and 316L SUS [5].
  A problem related with metallic materials used at high temperature is the intermetallic diffusion of the palladium in the metallic support [6–8]. In order to avoid this problem, it is necessary to realize a specific interdiffusion barrier [1] between the Pd-based selective layer and the metallic support.
- Owing to developments in the filtration industry, various ceramic filters with controlled pore size are commercially available today. These materials can also be used as supports for palladium composite membranes, but their poor physical strength and, in particular, their incompatibility with conventional techniques for joining parts (e.g., welding) causes difficulties in membrane module construction. The ceramic materials used for membranes syntheses reported in literature are: α-alumina, γ-alumina [9, 10] and yttria-stabilized zirconia (YSZ) [11] also in multilayer substrates $Pd/\gamma\text{-}Al_2O_3/\alpha\text{-}Al_2O$ and $Pd/YSZ/\alpha\text{-}Al_2O_3$ [11].

3.2.2 Interdiffusion Barrier

Such a barrier has to have a good chemical stability and has to be a reduced thickness (few μm) in order to allow the gas crossing.

Its unique rule is to avoid the Pd alloy interdiffusion in the steel support.

One of the main objective to aim in realizing an interdiffusion barrier is to obtain a layer extremely adherent, dense, homogeneous, and with continuous thickness.

Actually, several types of barrier materials have been tested in literature: TiN [6], TiO_2 [12], Al_2O_3 [8], $\alpha\text{-}Fe_2O_3$, $\gamma\text{-}Al_2O_3$, and YSZ [13]. TiN was considered the most promising type of barrier.

3.2.3 An Experimental Study: Support + Interdiffusion Barrier

In the frame of the mentioned research project, the Dept. of Chemical Engineering, Environment and Materials of the Rome University "La Sapienza", together with

Tecnimont-KT SpA, has focused on a traditional manufacturing path, realizing an innovative substrate for metallic membranes.

Two different configurations for the substrate have been considered:

- metallic tubular support with titanium nitride barrier;
- metallic planar support with alumina barrier.

To realize the support, sintered austenitic stainless steel (AISI316L) with a porosity >1 μm has been used. In order to obtain a suitable support with smaller pore size the AISI 316L has been resintered by GKN Sinter Metal Filters GmbH, Radevormwald, Germany. GKN technology consists in spraying the sintered stainless steel with a fine layer of nano-powder. The second sinterization step allows realizing a steel support with a surface porosity of 0.1 μm. Figures 3.2 and 3.3 show the support surface and section after resintering step, respectively.

The GKN supports have been coated with TiN as intermetallic diffusion barrier with three different thicknesses (0.6, 1.5, and 2.6 μm): the Titanium based layer was realized by Galileo Avionica SpA through Physical Vapor Deposition (PVD) technique. The 2.6 μm thickness layer has homogenously recovered the porous stainless steel (PSS) support, and has shown a very high permeance to hydrogen. Figure 3.4 shows the resintered support surface covered with a 2.6 μm layer of TiN. Figure 3.5 shows the section support.

It is highlighted the excellent adhesion of TiN on the low porosity support.

AISI316L resintered planar supports have been realized with the same GKN technology. This support has been covered by Acktar Ltd with a 5 μm Al_2O_3 barrier by RTE technology.

Figures 3.6 and 3.7 highlight the excellent coverage of the surface.

Two optimum substrates for final membrane synthesis have been identified.

Fig. 3.2 SEM analysis: AISI316L resintered support surface (×1000)

Fig. 3.3 SEM analysis: AISI316L resintered support section (×1000)

Fig. 3.4 Macrography: AISI316L resintered support surface coated with TiN 2.6 μm

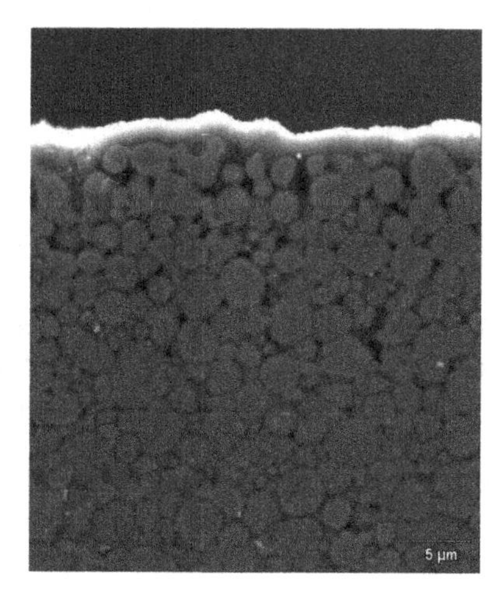

Fig. 3.5 SEM analysis: AISI316L resintered support section coated with TiN 2.6 μm (×2500)

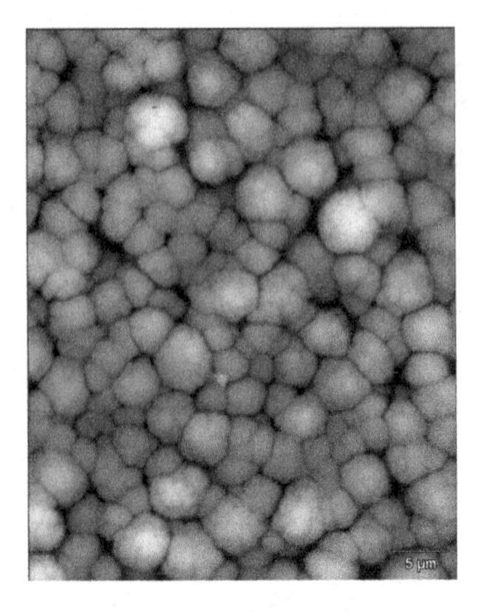

Fig. 3.6 SEM analysis: AISI316L resintered support surface coated with Al_2O_3 5 μm (×2500)

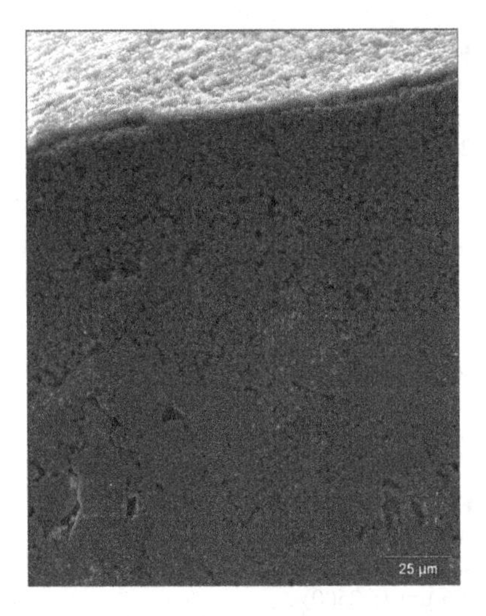

Fig. 3.7 SEM analysis: AISI316L resintered support section coated with Al_2O_3 5 μm (×600)

3.2.4 Pd Alloy Selective Layer

The selective layer, that allows separating the hydrogen from the other gases, is strictly connected with the chemical and physical membrane characteristics.

The morphology of the surface on which the selective layer has to be deposited is a key issue for a proper integration. Pore size and porosity of the deposition surface influence the membrane thickness and consequently its permeability: increasing the deposition surface pore size, the thickness of the alloy required to obtain a dense and continuous layer increases reducing hydrogen flow. Moreover, increasing porosity, the effective available area for hydrogen permeation increases together with hydrogen flow [14].

Another parameter that has to be considered is the support roughness: higher is the roughness of the deposition surface, more irregular is the deposition surface.

3.2.4.1 Permeation Mechanisms of Hydrogen in Palladium

Hydrogen-permeable metal membranes are extraordinarily selective, being extremely permeable to hydrogen but essentially impermeable to all other gases.

The gas transport mechanism is the key to the high selectivity. Hydrogen permeation through a metal membrane follows the multistep process illustrated in Fig. 3.8 and already described in Chap. 2. Hydrogen molecules from the feed gas are adsorbed on the membrane surface, where they dissociate into hydrogen atoms. Each individual hydrogen atom loses its electron to the metal lattice and diffuses through the lattice as an ion. Hydrogen atoms emerging at the permeate side of the membrane reassociate to form hydrogen molecules, then desorb, completing the permeation process. Only hydrogen is transported through the membrane by this mechanism; all other gases are excluded [15].

The palladium–hydrogen system is a two-component system (Pd and H) whose degrees of freedom are determined by the number of existing phases at working temperature and pressure.

The palladium–hydrogen system (Fig. 3.9) [16] shows the presence of two phases, α and α'. In both the phases, hydrogen occupies, randomly, the interstitial octahedral sites of the f.c.c. Pd lattice.

The α phase is a low hydrogen concentration phase and it can be seen as a solution of atomic hydrogen in palladium.

The α' phase is a high hydrogen concentration phase and it can be seen as an expanded phase (hydride). The α and α' phases are separated by a region where both coexist.

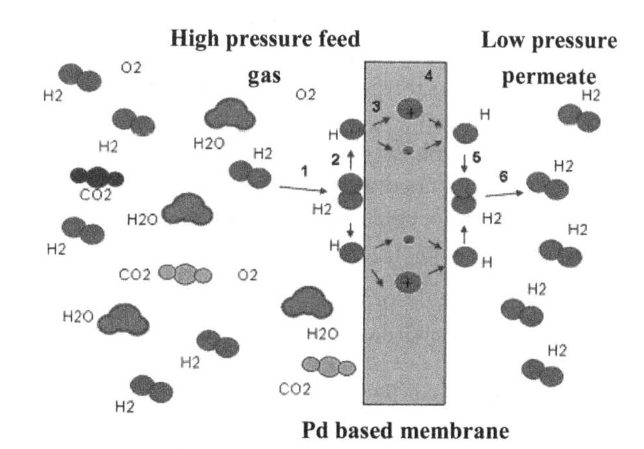

Fig. 3.8 Mechanism of permeation of hydrogen through metal membranes: (1) adsorption, (2) dissociation, (3) ionization, (4) diffusion, (5) recombination, (6) desorption

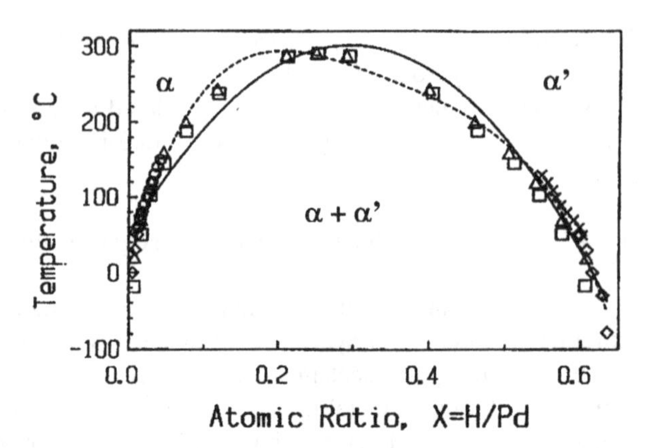

Fig. 3.9 T-X diagram for the Pd–H system with selected data. *Dashed line* gives a best fit (4th order polynomial) to the limit of the coexistence region as determined by the authors cited. *Solid line* represents the same data points fitted to a parabola (quadratic trimnomial): *open triangle* [17], *opentriangle*; [18], *cross symbol* [19], *open circle* [20], *open square* [21]

Above the critical temperature (293°C, 22.5 at %H) there is no distinction between the α and α' phase [16].

To draw the Pd–H phase diagram, we need to determine the absorption isotherms, indeed various authors, Wicke, Nernst, Frieske, Lasser, Blaurock [17–21], observed a hysteresis in the plateaus region, absent in monophasic regions, which translates into absorption process equilibrium pressure higher then desorption process equilibrium pressure (with the same concentration).

In the $\alpha \rightarrow \alpha'$ phase transition at temperatures and hydrogen pressures below 293°C and 2 MPa, respectively, the lattice expands at the phase transition point increasing in volume by about 10% [22]. This unit cell volume change can result in mechanical strains, physical distortions, and possibly failure of the palladium if cycled through the palladium hydride phase transition region.

Many authors tried to interpret the hysteresis phenomenon, including Ubbelohde [23]:

- when the α' hydride phase is formed, this phase is subjected to internal mechanical stress, so the free energy change is lower. When the desorption step occurs, hydrogen is preferentially released from regions subjected to stress. Hydrogen is so released at higher pressure than would occur from a stable α' phase;
- the formed α' phase is in a state of "disorder" (with hydrogen atoms in octahedral and tetrahedral interstitial positions) compared to the ordered "state" (hydrogen in octahedral sites) and the transition between the two states is delayed by an energy barrier [24].

So, during the desorption both the mechanical deformation that the disordered state would lack and the curve would be close (if not identical) to the equilibrium condition.

The problem can be remedied by exposing the palladium to hydrogen only at temperatures above 293°C. Since the hydride is the only phase present at these conditions, palladium will not be subjected to the stresses caused by the phase transition [25].

If the sorption and dissociation of hydrogen molecules is a rapid process, then the hydrogen atoms on the membrane surface are in equilibrium with the gas phase. The concentration c of hydrogen atoms on the metal surface is given by Sievert's law:

$$c = Kp^{1/2} \tag{3.1}$$

where K is Sievert's constant and p is the hydrogen pressure in the gas phase. At high temperatures (>300°C), the surface sorption and dissociation processes are fast, and the rate-controlling step is diffusion of atomic hydrogen through the metal lattice. Holleck et al. [26] have observed that the hydrogen flux through the metal membrane is proportional to the difference of the square roots of the hydrogen pressures on either side of the membrane. At lower temperatures, however, the sorption and dissociation of hydrogen on the membrane surface become the rate-controlling steps, and the permeation characteristics of the membrane deviate from Sieverts' law predictions [15].

Diffusion of the hydrogen through the palladium is attributed to the "jumping" of hydrogen atoms through the octahedral interstitial sites of the face-centered cubic palladium lattice [25, 27]. The lattice-diffusional mode of mass transfer for hydrogen results in the essentially infinite selectivity observed with dense palladium membranes. The hydrogen permeability of palladium increases with temperature because the endothermic activation energy for diffusion dominates the exothermic adsorption of hydrogen on palladium [25, 28].

Many disputes were created about the adsorption speed of hydrogen on the sample. Of course, as already Wagner noticed [29], the sample history (preparation method, purity, crystalline state, cleaning of the membrane surface by oxidation or sulfur deposition from trace amounts of hydrogen sulfide) dramatically affects the absorption mechanism.

Indeed, recent studies of Kay et al. [30, 31] showed that the probability of hydrogen molecule dissociation on the surface of palladium is close to unit. This circumstance excludes that this step is the slow step of the absorption process. By thermal desorption measurements, the above-mentioned authors argued that the phenomenon speed is governed by atomic hydrogen diffusion from the surface towards the massive phase.

At higher H pressure, the system no longer behaves ideally and the absorption initial speed increases with charge. This is interpreted as an effect due to lattice expansion which accompanied the absorption of hydrogen: indeed, the same effect of speed raising is achieved mechanically expanding the palladium lattice or introducing atoms of Ag (Pd–Ag alloy).

The silver, in addition to increasing the solubility of hydrogen in palladium, contributes to mechanical stability and lower cost of the membrane.

The optimal weight composition of the PdAg alloy is 23% Ag, 77% Pd [32, 33].

3.3 Membranes Deposition Techniques

Any thin film deposition process involves three main steps: (a) production of the appropriate atomic, molecular, or ionic species; (b) transport of these species to the substrate through a medium; (c) condensation on the substrate, either directly or via a chemical and/or electrochemical reaction, to form a solid deposit.

There are many processes today used for film deposition. However, often these are variants of two basic processes: physical process and chemical process.

Only the most widely used processes for the production of composite membranes for hydrogen separation are mentioned below.

3.3.1 Physical Processes

The physical process consists of physical vapor deposition (PVD) and it can be classified as follows: (1) thermal evaporation; (2) sputtering.

(1) Thermal evaporation: The thermal evaporation process comprises evaporating source materials in a vacuum chamber below 10^{-4} Pa and condensing the evaporated particles on a substrate.
The thermal evaporation processes are classified as [34]:

   (a) *Vacuum deposition* Resistive heating is most commonly used for the deposition of thin films. The source materials are evaporated by a resistively heated filament or boat, generally made of refractory metals such as W, Mo, or Ta, with or without ceramic coatings. Crucibles of quartz, graphite, alumina, beryllia, boron-nitride, or zirconia are used with indirect heating. The refractory metals are evaporated by electron-beam deposition since simple resistive heating cannot evaporate high melting point materials.
   (b) *Pulsed laser deposition (PLD)* This process is an improved thermal process used for the deposition of alloys and/or compounds with a controlled chemical composition. In laser deposition, a high-power pulsed laser is irradiated onto the target of source materials through a quartz window. A quartz lens is used to increase the energy density of the laser power on the target source. Atoms that are ablated or evaporated from the surface are collected on nearby sample surfaces to form thin films [35].

(2) Sputtering: When a solid surface is bombarded with energetic particles such as accelerated ions, surface atoms of the solid are scattered backward due to collisions between the surface atoms and the energetic particles. This phenomenon is called *back-sputtering,* or simply *sputtering*.

When a thin foil is bombarded with energetic particles, some of the scattered atoms transmit through the foil. The phenomenon is called *transmission sputtering* [36].

3.3.2 Chemical Processes

The chemical process can be classified as follows: (1) thermal chemical vapor deposition (CVD); (2) plating process.

(1) CVD process: The CVD process is realized when a volatile compound of the substance to be deposited is vaporized, and the vapor is thermally decomposed into atoms or molecules, and/or reacts with other gases, vapors, or liquids at the substrate surface to yield nonvolatile reaction products on the substrate. Most CVD processes operate at relative high temperature (near 1,000°C) in the pressure range of a few hundred Pa to above the atmospheric pressure of the reactants. Several CVD processes are proposed to increase the efficiency of the chemical reaction at lower substrate temperatures [34]. Plasma-assisted chemical vapor deposition (PACVD) is one of the modifications of the conventional CVD process. In the PACVD system, electric power is supplied to the reactor to generate the plasma. The ions in the plasma show slightly higher energy than the neutral gas molecules at room temperature. Typically the temperature of the ions in plasma is around 500 K.

(2) Plating processes:

   (a) *Electroplating* consists of the deposition of a metallic coating on an electrically conducting surface which acts as the cathode in an electrolytic cell, whose solution contains ions of the metal to be deposited. This is a relatively complex technology that involves a large number of steps [37].
   (b) *Electroless plating* is a non galvanic type of plating method that involves several simultaneous chemical reactions in an aqueous solution occurring without the use of external electric current. The chemical reactions are accomplished when hydrogen is released by a reducing agent, normally sodium hypophosphite, and is oxidized to produce a negative change on the surface of the substrate. This autocatalytic deposition method enables metal coating of non conductive textile material which can be used for precision work in conventional manufacturing. Unlike electroplating, the absence of electric field contributes to a uniform plating thickness.

Under properly controlled conditions all the above-mentioned methods produce good quality thin layers but electroless plating has the advantage of easy scale-up and the flexibility to coat the metal film on supports of different geometry.

However, the main disadvantage is the difficulty to control the composition of the alloy.

PVD sputtering has several advantages like:

(a) synthesis of ultrathin films with minimal impurity;
(b) easily controllable process parameters;
(c) flexibility for synthesizing alloys;
(d) the ability to generate nanostructured films.

The last two points are very important in membrane preparation for hydrogen separation because fabricating membrane alloys helps to overcome the problem of hydrogen embrittlement, while the nanostructured films may have unique size-dependent properties, e.g., a high hydrogen permeation [38].

3.4 Membranes Available on the Market

Only few companies are able to supply Pd-based membranes or membranes "modules"; market is still limited at laboratory scale membranes or modules for small pilot units. Here, below a short review of some membranes providers.

3.4.1 ECN Hydrogen Separation Modules (Hysep)

The Energy research Centre of the Netherlands (ECN), produces and offers a line of hydrogen separation modules (Hysep) on a pre-commercial basis for evaluation purposes. The technology is based on Palladium membranes which are capable of separating high purity hydrogen from a gas mixture.

An essential element of the Hysep® technology is the use of thin film palladium composite membranes to enable low cost and reliable hydrogen separation. The supported palladium layer in the Hysep® module has a thickness as low as 3–9 μm, a substantial improvement over current commercial available palladium membranes, which are based on self supporting metal foils with a thickness of 20–100 μm.

ECN has developed for Tecnimont-KT in the framework of FISR project described in Chap. 10 a module with 0.4 m^2 of a few μm palladium membranes with tubular geometry (Fig. 3.10).

Fig. 3.10 The Hysep1308 hydrogen separation module supplied for Tecnimont-KT pilot plant in Chieti

Fig. 3.11 MRT module

3.4.2 MRT Hydrogen Separation Modules

MRT is a Vancouver-based private company interested in hydrogen purifiers to provide high purity hydrogen and to recover hydrogen from mixed gas streams.

MRT produced membranes either as rolled foils or as deposited thin films (8–15 μm). In addition, patented bonding techniques have been developed to permanently attach membranes to support modules with a perfect, hydrogen-tight seal.

For membranes thinner than 15 μm, MRT uses a proprietary coating technique. Prototype membranes as thin as 8 μm, tested by MRT, have been produced and show excellent performance and longevity.

The MRT Purifier, developed for Tecnimont-KT in the framework of FISR project, has been designed to house five membrane modules and operate at 450°C and up to 25 barg. Each module consists of two double sided, planar 30 cm × 12 cm membrane panels welded in series. Each panel has a palladium (Pd) alloy active membrane area of 0.03 m^2 per side for a total installed membrane area of 0.6 m^2 in the purifier. The modules are housed in a rectangular core which, along with the inlet distributor, promotes uniform reformate flow across the membrane modules. The core assembly (Fig. 3.11) is housed inside a pressure vessel.

3.4.3 Hydrogen Selective Membranes Produced in Japan

An important Japanese Company (JC) is developing a gas separation membrane, which efficiently recovers hydrogen, by forming a film on a porous ceramic substrate using palladium alloy known for its feature of selective permeation of hydrogen. The key is to simultaneously achieve cost effectiveness and high hydrogen selectivity, by making expensive palladium membranes thinner.

At laboratory scale, the JC has realized Pd–Ag membranes on ceramic substrates.

Fig. 3.12 Hydrogen selective membranes produced in Japan

The membranes are produced in a three-step procedure: at first, Pd is deposited onto the Al_2O_3 support by electroless plating technique, and then Ag is layered on by electroplating using the Pd layer as electrode. The layered Pd–Ag membrane is finally heat-treated to obtain the Pd–Ag alloy membrane.

The resulting membranes are tubular with an external diameter of about 1.0 cm, an effective length of about 9.0 cm and a Pd–Ag coating deposited on the external surface with a selective layer of about 28.3 cm^2.

JC has realized for Tecnimont-KT three tubular Pd–Ag membranes with a total surface area of 0.12 m^2 (Fig. 3.12).

3.4.4 SINTEF Hydrogen Selective Membranes

SINTEF, research and educational centre in the field of environmentally technology, has developed a technique for the manufacturing of palladium-based hydrogen separation membranes based on a two-step process allowing a reduction of membrane thickness making palladium membranes economically viable. First, a defect-free Pd-alloy thin film is prepared by magnetron sputtering onto the 'perfect surface' of a silicon wafer. In a second step, the film is removed from the wafer. These films may subsequently either be used self-supported or integrated with various supports of different pore size, geometry, and size (Fig. 3.13). This allows, for example, the preparation of very thin (approximately 2–3 μm) high-flux membranes supported on macroporous substrates, which can operate at high pressures. The efficiency of these membranes has been investigated by SINTEF at elevated temperatures and pressures and reported in a remarkable number of publications on scientific journals.

In pure H_2, applying a H_2 feed pressure of 26 bars, one of the highest H_2 fluxes reported, 2477 ml cm^{-2} min^{-1} (STP) or 6.1×10^4 kg H_2 m^{-2} h^{-1}, which corresponds to a permeance of 1.5×10^{-2} mol m^{-2} s^{-1} $Pa^{-0.5}$, was measured at 400°C. In water gas shift (WGS) conditions (57.5% H_2, 18.7% CO_2, 3.8% CO, 1.2% CH_4 and 18.7% steam), SINTEF membranes has shown a H_2 permeance of

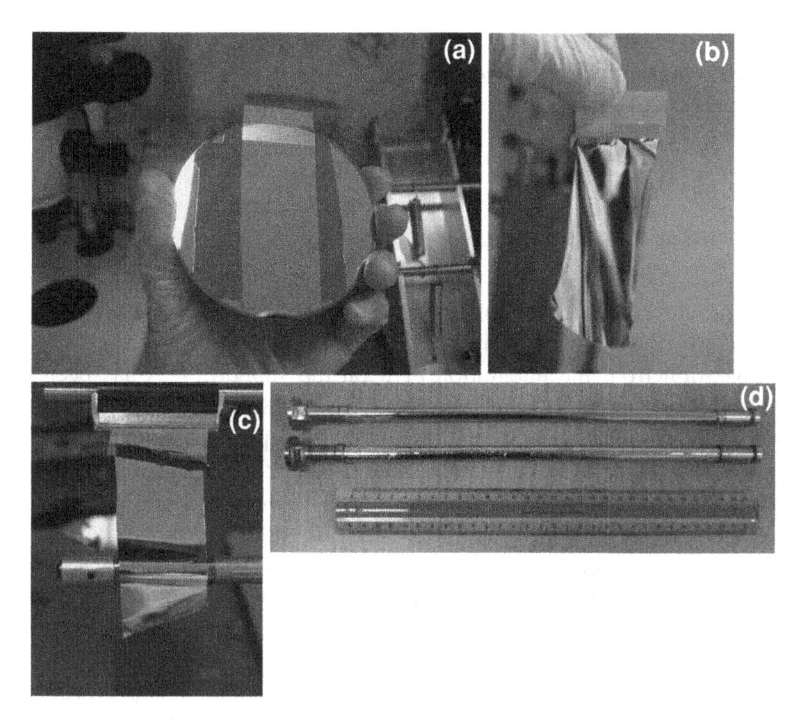

Fig. 3.13 SINTEF membranes manufacturing process; **a** Pd-alloy film prior to removal from the silicon wafer; **b** unsupported Pd-alloy film; **c** Pd-alloy film during wrapping on the tubular porous stainless steel (PSS®) support; **d** Composite Pd-alloy/stainless steel composite membranes (CACHET-public workshop, Athens 24.04.2007)

1.1×10^{-3} mol m^{-2} s^{-1} Pa$^{-0.5}$ at 400°C and 26 bar feed pressure. Operating the membrane for more than 1 year under various conditions (WGS and $H_2 + N_2$ mixtures) at 10 bars indicated no membrane failure. A membrane life time of several (2–3) years ($T \leq 425$°C) is assessed for the employed experimental conditions based on these long-term stability tests. Post-process characterization showed a considerable grain growth and of micro-strain relaxation in the Pd–23%Ag membrane after the prolonged permeation experiment. Changes in surface area are relatively small.

3.5 Membrane Manufacturing Strategy

To lower the production costs, it is important to develop a proper membranes manufacturing strategy (MMS) which involves two main aspects:

- the manufacturing process itself which will give the business a distinct advantage in the market-place through unique technology for instance;

- manufacturing associated activities in terms of infrastructure design as controls, procedures, subcontractors selection and so on that are involved in the main process aspects of manufacturing.

3.5.1 Manufacturing Process

Based on our current understanding and experience in composite membrane, one way to simplify the manufacturing process is to separate the Pd-based selective layer preparation from the integration of it on the support. This has been demonstrated by Bredesen and Klette [39] and shown in Sect. 3.4.4.

Moving ahead such a concept, the manufacturing process should consist of a batch process on three independent steps. The choice of a batch process is a logical one because it provides similar items on a repeat basis, usually in larger volume. Batch procedure divides the manufacturing task into a series of appropriate operations, which together will make the product involved. It is then not so difficult to define the main steps in such a process: first of all, the Pd-alloy selective layer preparation, and secondly, the support preparation and finally the membrane module assembling and testing.

One way to approach the membrane and support fabrication is to consider two distinct lines, each one delivering the distinctive product, which at the end of the process is tested for quality control before moving to the integration.

A roll-in process as proposed by Acktar, and detailed in Fig. 3.14, based on vapor deposition could produce the thin Pd-alloy layer, meanwhile planar or

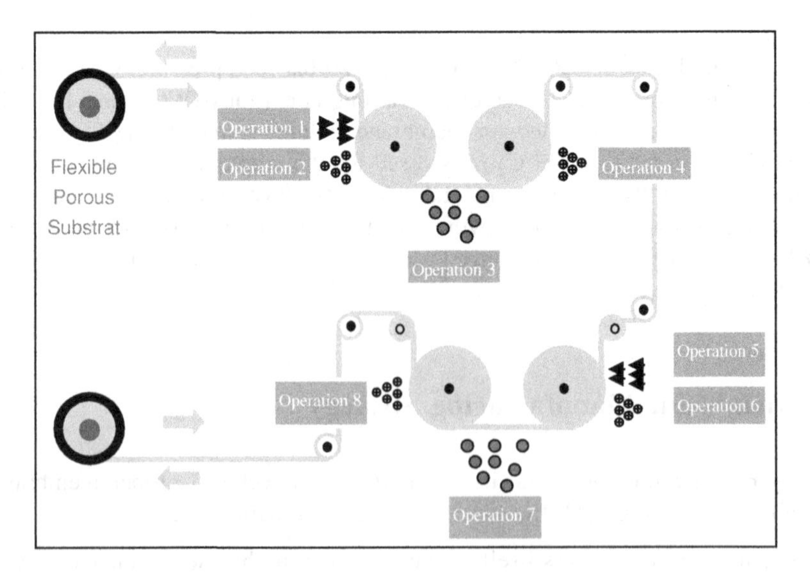

Fig. 3.14 Scheme of a roll-in process based on vapor deposition (courtesy form "Acktar Ltd.")

tubular PSS support can be bought on the market by a proper subcontractor and coated with a fine porous as metallic inter-diffusion barrier in order to realize the membrane substrate.

Once the two single specific components have been prepared and tested, the membrane can be formed by laying the membrane on the support. Several membranes can be assembled together to build a module.

Because of the metal structure of the support, the proposed composite membranes provide a solution to the problems that result from high welding temperatures or high mechanical compressing force caused by the joining of a composite membrane with other parts through Swagelok, welding, brazing, and gasket, etc.

A flat configuration of the membranes provides solutions to the problems associated with module assembling. For planar porous metal substrate, a solid frame, useful for realizing the module sealing, can be welded to its perimeter.

3.5.2 Manufacturing-Associated Activities

The manufacturing task of choosing a process and necessary hardware is not simply. Process choice concerns the features for hardware, the tangible ways in which the products are manufactured; but the task is more than this.

The associated structures, controls, procedures, and other systems within manufacturing are equal necessary for successful, competitive manufacturing performance.

In our MMS, quality controls for instance related to the thin films but also to the support and the membrane overall, constitute an essential part in the manufacturing task. Creating a quality function in the organization to supervise such operational controls is an essential key issue to develop and manage.

Fail to develop a proper infrastructure on the complex process of making membranes may result in the impossibility to reach the target costs.

3.6 Membrane Costs Analysis and Dynamics

In order to tackle this issue and to be able to forecast a production cost for thin Pd-based membranes, it is important to introduce the concept of "economics of learning" in understanding the behaviour of all added costs of membranes as cumulative production volume increased. Such economics of learning or law of the experience may be expressed more precisely in an algebraic form:

$$c_n = c_1 n^{-a} \tag{3.2}$$

where c_1 is the cost of the unit production (square meter of membrane for instance), c_n is the cost of the nth unit of production, n is the cumulative volume of production, and a is the elasticity of cost with regard to output.

Graphically, the experience curve is characterized by a progressively declining gradient, which, when translated into logarithms, is linear.

The size of experience effect is measured by the proportion by which costs are reduced with subsequent doublings of aggregate production.

3.6.1 Plotting the Experience Curve for the Membrane Production

Constructing an experience curve is a simple matter once the data are available. Of course for the Pd-based or ceramic membrane such dates are limited to minimal surface (less than 1 m^2), which can, however, be used as starting point of the curve. The other issue associated with drawing an experience curve is that cost and production data must be related to a "standard product", which is not the case due to the fact that in the membrane technology no standard is yet emerged and there is a lot of discussion on the membrane composition and preparation method, supporting matrix and other mechanical and construction details.

It is, however, a fact that costs decline systematically with increases in cumulative output.

The assumptions made in the following are:

$$c_1 = 50{,}000\,€$$

$$a = 0.25$$

where c_1 value derived by Tecnimont-KT recent experience in building a pilot unit (refer to Chap. 10), meanwhile the "a" factor was assumed as average value typically between 20 and 30%.

Using such a data is possible to forecast the cost for m^2 of membrane module versus the cumulative value of production, expressed in terms of m^2.

Table 3.1 and Fig. 3.15 show such data.

It is important to note that the cost of bought-in materials and components for a few micron Pd membranes is around 700 € as expressed by ECN [40], and such information fits reasonably well on what was presented and indicated by some authors [41].

Table 3.1 Cost for m^2 of membrane module versus cumulated production

Cumulated production m^2	€ cost per m^2
1,000	8,900
10,000	5,000
100,000	2,800
1,000,000	1,600
10,000,000	900

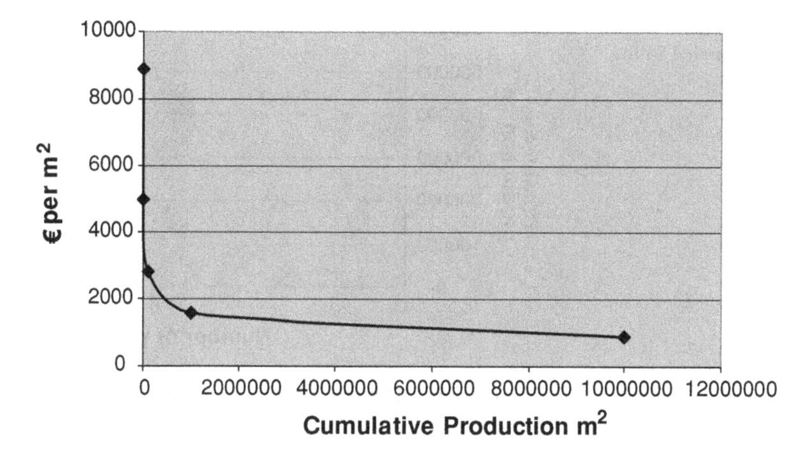

Fig. 3.15 Cost for m^2 of membrane module versus cumulated production

3.6.2 Strategy Implication of the Experience Curve

From the drawn experience curve, some implications for the membranes market business strategy can be extracted.

The first and more important question to answer is when a 1,000,000 m^2 of membrane module cumulative production could be reached in order to have a unit cost around 1.600 € per m^2 of membrane.

In order to answer such a question, further considerations need to be developed, to relate surface to membrane module to the H_2 production and to the introduction of such a new technology in the market.

On previous published data, Iaquaniello [42] was calculating for a open membrane reactor architecture a surface of 1,000 m^2 for an installed capacity of 10,000 Nm^3/h of Hydrogen. The envisaged installed capacity in the hydrogen market is today around 1 MM Nm^3/h of capacity per year, which translated into a production of 100,000 m^2 of membrane year, once the new technology will supersede the conventional one.

To derive the rate of membranes technology introduction in the market a Volterra equation was considered:

$$x = A \Big/ \left(1 + e^{(Bx)}\right) + C \tag{3.3}$$

where A, B, C are constants and x is the cumulative production.

Such equation, also called "S logistic curve" is used to describe a process with a low growth which accelerate with time to seem an exponential growth. A 10-year period (2012–2022) is considered to achieve 50% substitution in the conventional market starting from 2012, which roughly implies that over the next decade half a million of square meters of membranes modules could be produced. With such cumulative production around year 2020, the membrane cost per m^2 could reach

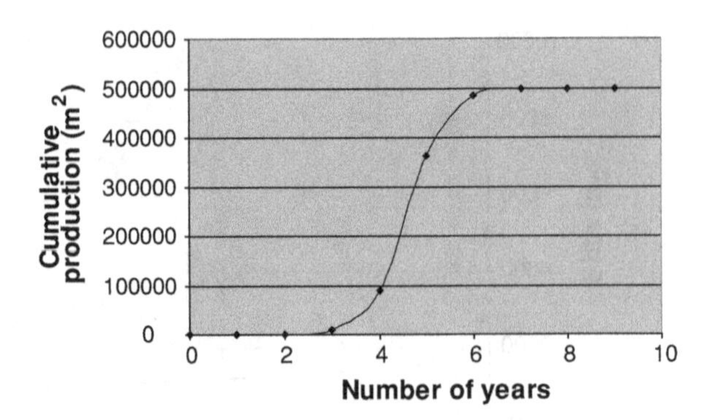

Fig. 3.16 Cumulative production coupled to the "*S*" curve

the target of 1.600 € per m^2 and the overall market will have a size of 1 billion of € per year.

The Fig. 3.16 represents the cumulative production coupled to the "S" curve.

The approach used to determine the growth of the membranes market, together with the cumulative production does not, however, identify the real factors that determine its dynamics.

As matter of fact, the experience curve combines four sources of costs reduction: learning, economics of scale, process innovation, and improved production design.

Economics of scale, conventionally associated with manufacturing operations, is probably the most important of these costs drivers and exists wherever as the scale of production increases unit costs fall. A plant capacity has then an economic sense if a minimum efficiency plant capacity is reached.

This will imply that to reach the required reduction in the membrane cost, not only a few specialized technologies must emerge, but the production market will be concentrated in few highly specialized production plants.

3.7 Conclusion

Only few companies are able to supply Pd-based membranes or membranes "modules"; market is still limited at laboratory scale membranes or modules for small pilot units.

A reasonable price within 1,500–2,000 € for m^2 of thin Pd based membranes and even less for Si or inorganic membranes could be reached at the end of next decade only if a cumulative volume of production high enough is reached.

A "S" logistic curve was used to describe the substitution process of the conventional technology.

A precondition for such behavior is the emerging of one or two technologies which can sustain costs reduction based on economics of scale.

Physical vapor deposition technology to produce membranes on a roll-in process and a metallic support is indicated by the authors as one of the more promising technology for such industrial mass production.

References

1. Li A, Grace JR, Lim CJ (2007) Preparation of thin Pd-based composite membrane on planar metallic substrate Part II. Preparation of membranes by electroless plating and characterization. J Membr Sci 306:159–165
2. Tong JH, Kashima Y, Shirai R, Suda H, Matsumura Y (2005) Thin defect-free Pd membrane deposited on asymmetric porous stainless steel substrate. Ind Eng Chem Res 44:8025–8032
3. Checchetto R, Bazzanella N, Patton B, Miotello A (2004) Palladium membranes prepared by r.f. magnetron sputtering for hydrogen purification. Surf Coat Tech 177–178:73–79
4. Chen SC, Tu GC, Hung CCY, Huang CA, Rei MH (2008) Preparation of palladium membrane by electroplating on AISI 316L porous stainless steel supports and its use for methanol steam reformer. J Membr Sci 314:5–14
5. Nam SE, Lee SH, Lee KH (1999) Preparation of a palladium alloy composite membrane supported in a porous stainless steel by vacuum electrodeposition. J Membr Sci 153:163–173
6. Shu J, Adnot A, Grandjean BPA, Kaliaguine S (1996) Structurally stable composite Pd-Ag alloy membranes: introduction of a diffusion barrier. Thin Solid Films 286:72–79
7. Wang D, Tong HH, Xu HY, Matsumura Y (2004) Preparation of palladium membrane over porous stainless steel tube modified with zirconium oxide. Catal Today 93–95:689–693
8. Yepes D, Cornaglia LM, Irusta S, Lombardo EA (2006) Different oxides used as diffusion barriers in composite hydrogen permeable membranes. J Membr Sci 274:92–101
9. ZHANG K, Gao H, Rui Z, Lin Y, Li Y (2007) Preparation of thin palladium composite membranes and application to hydrogen/nitrogen separation. Chin J Chern Eng 15(5):643–647
10. Yan S, Maeda H, Kusakabe K, Morooka S (1994) Thin palladium membrane formed in support pores by metal-organic chemical vapor deposition method and application to hydrogen separation. Ind Eng Chem Res 33:616–622
11. Zhang K, Wei X, Rui Z, Li Y, Lin YS (2009) Effect of metal-support interface on hydrogen permeation through palladium membranes. AIChE J 55(3):630–639
12. Huang Y, Dittmeyer R (2006) Preparation and characterization of composite palladium membranes on sinter-metal supports with a ceramic barrier against intermetallic diffusion. J Membr Sci 282:296–310
13. Zhang K, Gao H, Rui Z, Liu P, Yongdan Li, Lin YS (2009) High temperature stability of palladium membranes on porous metal supports with different intermediate layers. Ind Eng Chem Res 48:1880–1886
14. Mardilovich IP, Engwall E, Ma YH (2002) Dependence of hydrogen flux on the pore size and plating surface topology of asymmetric Pd-porous stainless steel membranes. Desalination 144:85–89
15. Baker RW (2004) Membrane technology and applications. Wiley, New York
16. Manchester FD, San-Martin A, Pitre JM (2000) H-Pd (hydrogen-palladium). In: Manchester FD (ed) Phase diagrams of binary hydrogen alloy. ASM International, Metals Park, pp 158–181
17. Wicke E, Nernst GH (1964) Phase diagram and thermodynamic behaviour of the palladium-hydrogen and of the palladium-deuterium system at normal temperature; H/D separation effect. Ber Bunsenges Phys Chem 68:224–235
18. Frieske H, Wicke E (1973) Magnetic susceptibility and equilibrium diagram of PdHn. Ber Bunsenges Phys Chem 77(1):48–52
19. Lasser R, Klatt KH (1983) Solubility of hydrogen isotopes in palladium. Phys Rev B 28:748–758
20. Lasser R (1985) Isotope dependence of phase boundaries in PdH, PdD, and PdT systems. J Phys Chem Solids 46:33–37
21. Wicke E, Blaurock J (1987) New experiments on and interpretation of hysteresis effects of Pd-D2 and Pd-H2. J Less-Common Met 130:351–363

22. Grashoff GJ, Pilkington CE, Corti CW (1983) The purification of hydrogen—a review of the technology emphasizing the current status of palladium membrane diffusion. Plat Met Rev 27(4):157–169
23. Ubbelohde AR (1937) Some properties of the metallic state I—metallic hydrogen and its alloys. Proc R Soc Lond A 159:295–306
24. Bragg WL, Williams EJ (1934) Effect of thermal agitation on atomic arrangement in alloys. Proc R Soc Lond A 145:699–730
25. Morreale BD, Ciocco MV, Enick RM, Morsi BI, Howard BH, Cugini AV, Rothenberger KS (2003) The permeability of hydrogen in bulk palladium at elevated temperatures and pressures. J Membr Sci 212:87–97
26. Holleck GL (1970) Diffusion and solubility of hydrogen in palladium and palladium silver alloys. J Phys Chem 74(3):503–511
27. Volkl J, Alefeld G (1975) Hydrogen diffusion in metals. In: Nowick AS, Burton JJ (eds) Diffusion in solids, recent developments. Academic Press, New York, pp 232–295
28. Buxbaum RE, Kinney AB (1996) Hydrogen transport through tubular membranes of palladium-coated tantalum and niobium. Ind Eng Chem Res 35:530–537
29. Wagner C (1932) Kinetics of reaction H_2 (gas) _ 2H(dissolved in palladium). Zeitschrift Phys Chem A159:459–469
30. Kay BD, Peden CHF, Goodman DW (1986) Kinetics of hydrogen absorption by Pd(110). Phys Rev B 34(2):817–822
31. Kay BD, Peden CHF, Goodman DW (1986) Kinetics of hydrogen absorption by chemically modified Pd(110). Surf Sci 175(1):215–225
32. Arstad WMB, Klette H, Walmsley JC, Bredesen R, Venvik H, Holmestad R (2008) Microstructural characterization of self-supported 1.6 μm Pd/Ag membranes. J Membr Sci 310:337–348
33. Jayaraman V, Lin YS (1995) Synthesis and hydrogen permeation properties of ultrathin palladium–silver alloy membranes. J Membr Sci 104:251–262
34. Wasa K, Kitabatake M, Adachi H (2004) Thin film materials technology: sputtering of compound materials. William Andrew Inc., New York
35. Koga T (1994) Off-axis pulsed laser deposition of YBaCuO superconducting thin films. MS Thesis, Royal Institute of Technology, Stockholm, Sweden
36. McClanahan D, Laegreid N (1991) Production of thin films by controlled deposition of sputtered material. In: Behrisch R, Wittmaack K (eds) Sputtering by particle bombardment III. Topics in applied physics, vol 64, Springer Verlag, Berlin, p. 339
37. Almeida E (2001) Surface treatments and coatings for metals. A general overview. 2. Coatings: Application processes, environmental conditions during painting and drying, and new tendencies. Ind Eng Chem Res 40:15–20
38. Pierson HO (1999) Handbook of chemical vapour deposition (CVD). Principles, technology, and applications, 2nd edn. Noyes Publications/William Andrew Publishing, New York
39. Bredesen R, Klette H (2000) US patent 6.086.729
40. Van Delft YC, Correia LA et al. (2007) Palladium membrane reactors for large scale production of hydrogen. 8th international conference of catalysis in membrane reactors, December 18-21 Kolkata (India)
41. U.S. DOE (2004) Hydrogen separation—technical targets. Office of Fossil Energy Hydrogen from Coal RD&D Plan, June 10, 2004
42. Iaquaniello G, Giacobbe F, Morico B, Cosenza S, Farace A (2008) Membrane reforming in converting natural gas to hydrogen: production costs, part II. Int J Hydrogen Energy 33: 6595–6601

Chapter 4
Membrane Reactors Modeling

Marcello De Falco

List of Symbols

a	External particles total surface to reactor volume ratio
A_i	Bed size heat exchanging surface
A_o	Heating fluid media heat exchanging surface
A_m	Log mean of A_i and A_o
B_H	Hydrogen permeability
C_A	Concentration of component A
$C_{A,s}$	Concentration inside catalyst particle
C_A^S	A-component concentration on catalyst surface
c_H	Hydrogen concentration
$c_{H_2,high}$	Hydrogen concentrations in high pressure mixture streams
$c_{H_2,low}$	Hydrogen concentrations in low pressure mixture streams
c_p	Gas mixture specific heat
$c_{p,perm}$	Specific heat in permeation zone
D	Hydrogen diffusion coefficient through membrane
D_e	Effective diffusivity in catalyst
D_{ea}	Effective axial diffusivity
D_{er}	Effective radial diffusivity
d_{mem}	Membrane diameter
$d_{mem,i}$	Internal membrane diameter
$d_{mem,o}$	External membrane diameter
d_p	Equivalent catalyst particle diameter
d_t	Internal tubular reactor diameter
E_a	Membrane permeability apparent activation energy
f	Friction factor

M. De Falco (✉)
Faculty of Engineering, University Campus Bio-Medico of Rome,
via Alvaro del Portillo 21, 00128 Rome, Italy
e-mail: m.defalco@unicampus.it

M. De Falco et al. (eds.), *Membrane Reactors for Hydrogen Production Processes*,
DOI: 10.1007/978-0-85729-151-6_4, © Springer-Verlag London Limited 2011

G	Superficial mass flow velocity
$(-\Delta H)$	Heat of reaction
h_{ex}	Heat transport coefficient in the external side
h_f	Heat transport coefficient between gas and solid phase
h_{perm}	Heat convective transport coefficient in the permeation zone
h_W	Heat transport coefficient in the first layer near the tube wall
J	Hydrogen flux through membrane lattice
J_{H_2}	Hydrogen flux through the membrane
K_g	Mass transport coefficient between gas and solid phase
k_{met}	Tube wall conductivity
P	Reaction pressure
P_0	Membrane permeability pre-exponential factor
Pe_a	Axial Peclet number
Pe_{mr}	Mass effective radial Peclet number
p_{H_2}	Hydrogen partial pressure
$p_{H_2,r}$	Hydrogen partial pressures in retentate side (high hydrogen partial pressure)
$p_{H_2,p}$	Hydrogen partial pressures in permeate side (low hydrogen partial pressure)
R	Universal gas constant
r_A	A-component reaction rate
r_{H_2}	Hydrogen reaction rate
r_i	*i*-Component reaction rate
R_{mem}	Selective membrane radius
S	Hydrogen solubility
T	Reactor operating temperature
T_{perm}	Permeation zone temperature
T_r	Temperature of heating/cooling fluid
T_{reac}	Reaction zone temperature
T^S	Temperature on catalyst surface
T_s	Temperature inside catalyst particle
T_W	Reactor tube wall temperature
U	Overall heat transport coefficient between the external and reaction bed
U_1	Overall heat transfer coefficient between reaction and permeation zone
u_s	Gas superficial velocity
$u_{s,high}$	Superficial gas velocity in separator high pressure zone
$u_{s,low}$	Superficial gas velocity in separator low pressure zone
$u_{s,perm}$	Superficial gas velocity in permeation zone
V_p	Pellet volume
z	Reactor axial coordinate
z_{perm}	Axial coordinate of membrane module

Greek Letters

α_{mem} Membrane thermal conductivity
ε Void fraction of the packed bed
δ Membrane thickness
η Catalyst effectiveness factor
λ_e Thermal conductivity in catalyst
λ_{er} Effective radial thermal conductivity
ξ Radial coordinate inside catalyst particle
μ_g Gas mixture viscosity
ρ_B Catalytic bed density
ρ_g Gas molar density
$\rho_{g,perm}$ Gas density in permeation zone

Abbreviations

BVP Boundary value problem
IMR Integrated membrane reactor
ODE Ordinary differential equation
PDE Partial differential equation
SMR Staged membrane reactor

4.1 Reactors and Separators Modeling Strategies

Throughout the ages, engineers and researchers have understood the importance of modeling and have faced the challenge of developing more and more accurate algorithms. Thanks to the strong development of both software and hardware, the computational loads of mathematical modeling have been increased in the years and nowadays the behavior of reactors and all chemical and physical phenomena occurring inside reaction environment can be evaluated in detail.

Practically, mathematical models are based on the conservation laws of mass, energy and momentum, which lead to mass, energy and momentum balances. The balances, together with transport and kinetics equations, form a set of equations (ODE or PDE) whose solution gives the component concentrations, temperature and pressure profiles inside the reactor. Mass and heat transport coefficients, reactants and products physical properties, catalyst efficiency factor and all parameters appearing in model equations have to be expressed.

In a rigorous model formulation, radial and axial mixing should be taken into account due to radial and gradients of compositions, pressure and temperature. As an example, if the elementary reaction

$$A \rightarrow B \tag{4.1}$$

is carried out in a tubular fixed bed reactor, the concentration of reactant A decreases along the reactor as shown in Fig. 4.1. The concentration drop is due to the reaction but at the same time the gradient associated to this profile leads to a mass flux in the direction shown in the figure. This flux would reduce the gradient itself and its entity should be taken in consideration in reactor modeling.

Moreover, catalysts are typically porous and the concentration profile of reaction components inside the particles, shown in Fig. 4.2, should be taken into account.

It is clear that a rigorous formulation of all phenomena occurring inside a reaction environment leads to a level of complexity difficult to be managed. The ability of reactor designers is mainly in introducing proper assumptions which reduce the formulation complexity but at the same time do not lead to unacceptable approximations.

Therefore, various models types can be developed and applied in each specific case.

In the following section, a classification of fixed bed reactor models is proposed and a brief description of each model type is reported.

4.1.1 Fixed Bed Reactor Model Classification

Mass, energy, and momentum local balances on a control volume are the basis of each model. In cylindrical symmetry, as for tubular reactors, the control volume is selected as shown in Fig. 4.3.

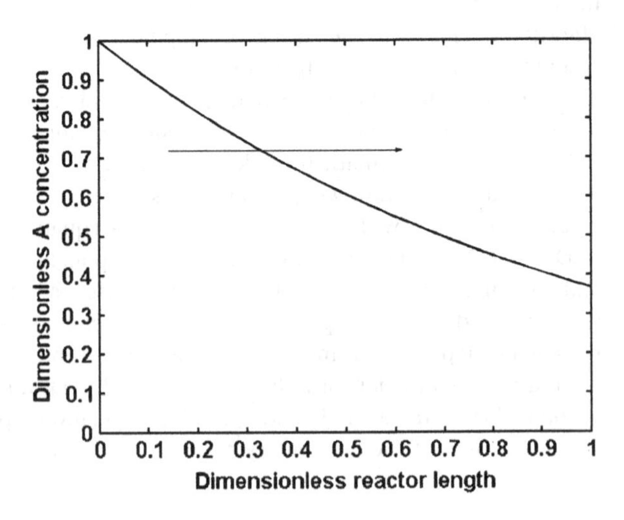

Fig. 4.1 Typical reactant profile in a tubular fixed bed reactor

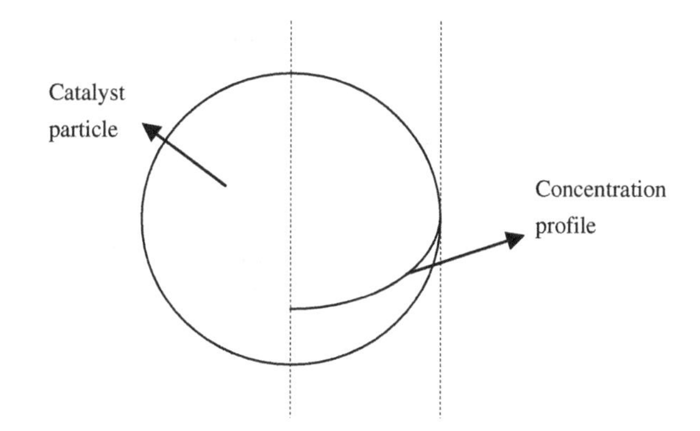

Fig. 4.2 Concentration profile inside a catalyst particle

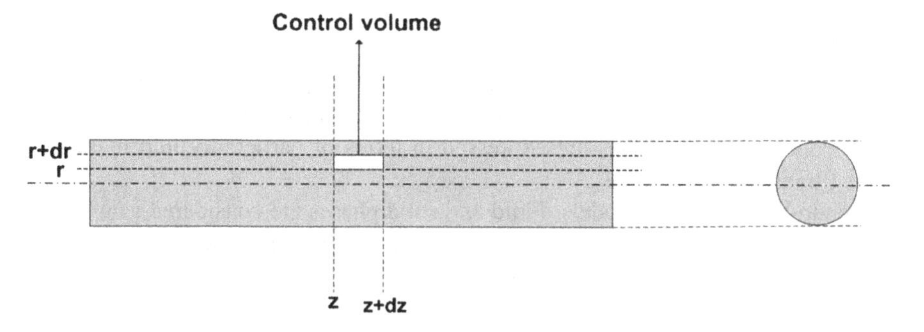

Fig. 4.3 Control volume definition for cylindrical symmetry

Balances for a generic variable X (component mass, energy, momentum) are expressed as follows:

$$\begin{aligned}&\text{Input rate of } X \text{ through control volume surface}\\&\quad - \text{Output rate of } X \text{ through control volume surface}\\&= \text{Accumulation rate of } X \text{ inside control volume}\\&\quad - \text{Generation rate of } X \text{ inside control volume}\end{aligned}$$

Input and output rates can be expressed as convective and diffusive terms, where convective terms are always associated to a mass flow while diffusive terms are associated to molecules fluxes. In reactors, X generation is due to reactions inside the volume, whose effects are:

- increase of products composition,
- reduction of reactants composition,
- increase or reduction of reactor temperature due to reaction exothermicity or endothermicity, respectively, and
- variation of fluid volume in dependence on algebraic sum of reaction stoichiometric coefficients.

Table 4.1 Pseudo-homogeneous and heterogeneous fixed bed reactor classification

Pseudo-homogeneous models	
	Ideal
One-dimensional	+ Axial mixing
Two-dimensional	+ Radial mixing
Heterogeneous models	
	Interfacial gradients
One-dimensional	+ Intraparticle gradients
Two-dimensional	+ Radial mixing

Accumulation term takes into account variations with time of reactor conditions.

Fixed bed reactors are systems composed by two or more physical phases since a fluid phase reacts over a solid catalyst. Models can be grouped in two broad categories:

- Heterogeneous models, by which the fluid and solid phases are modeled separately, imposing balance equations for each phase. Mass and heat fluxes between the solid and fluid phases are expressed in terms of particle-to-fluid mass and heat transport coefficients.
- Pseudo-homogeneous models. Fluid and solid phases are considered as a single pseudo-phase and the balances are imposed for only one phase. Heat and mass transport coefficients inside the bed are calculated by expressions which account for the simultaneous presence of two phases.

According to Froment–Bischoff [1] for each category, models can be classified in order of their growing complexity, as reported in Table 4.1.

In the following, a survey of model typologies is reported starting from the simplest ones and removing assumptions gradually to formulate more and more complex models. Only steady-state conditions are taken into account. For a deeper models description, please refer to [1–3].

4.1.1.1 Pseudo-Homogeneous Models

By pseudo-homogeneous models, fixed bed reactors are described without considering the existence of two different phases at least, but imposing balance equations for one single pseudo-phase and calculating chemical-physical properties and transport coefficients by applying empirical expressions.

The pseudo-homogeneous assumption reduces the complexity of the model implementation and resolution, but of course this model category gives less accurate results with respect to heterogeneous one.

The simplest fixed bed reactor model formulation is the *pseudo-homogeneous ideal model*, by which:

1. ideal plug-flow is imposed, assuming that compositions, temperature and pressure vary only in the axial direction;

2. axial mixing is neglected, and the only transport mechanism is the axial convective flux.

Assuming the elementary reaction (4.1) in gaseous phase, the mass balance equation for the ideal model is:

$$-u_s \frac{dC_A}{dz} = \rho_B r_A \tag{4.2}$$

where z is the reactor axial coordinate, u_s is the gas superficial velocity, assumed to be constant along z since the reaction (4.1) occurs without mole variation and neglecting the effects of temperature and pressure drop, C_A is the concentration of component A, ρ_B the catalytic bed density, and r_A the A-component reaction rate. The term $-u_s \frac{dC_A}{dz}$ represents the axial convective flux, while $\rho_B r_A$ is the generative term associated to the reaction.

Energy balance has the following form:

$$u_s \rho_g c_p \frac{dT}{dz} = (-\Delta H)\rho_B r_A - 4\frac{U}{d_t}(T - T_r) \tag{4.3}$$

where T is the reactor operating temperature, ρ_g the gas molar density, c_p the gas mixture specific heat, $(-\Delta H)$ the heat of reaction, T_r the temperature of heating/cooling fluid, U the overall heat transport coefficient between the external and reaction bed, d_t the internal tubular reactor diameter. The term $u_s \rho_g c_p \frac{dT}{dz}$ is heat convective flux, $(-\Delta H)\rho_B r_A$ is the generative term associated to the reaction and $4\frac{U}{d_t}(T - T_r)$ is heat exchanging flux between the reaction zone and the external.

The global heat transport coefficient U is calculated from the sum of different heat resistances in a series. For a packed bed reactor, the heat transport resistances concern:

- the heat medium side;
- the tube wall;
- the first layer of gas mixture, in which the heat transport only occurs by molecular conduction [4];
- the pseudo-homogeneous phase (gas + solid phases).

The expression for the calculation of U is:

$$U = \left(\frac{1}{h_{ex}} \cdot \frac{A_i}{A_o} + \frac{t}{k_{met}} \cdot \frac{A_i}{A_m} + \frac{1}{h_W} + \frac{d_t}{8\lambda_{er}} \right)^{-1} \tag{4.4}$$

where h_{ex}, k_{met}, h_W, and λ_{er} are the heat transport coefficient in the external thermal fluid, the tube wall conductivity, the heat transport coefficient in the first layer near the tube wall and the effective radial thermal conductivity of the pseudo-homogeneous phase, respectively, while A_i, A_o, and A_m are the bed size heat exchanging surface, the thermal fluid heat exchanging surface and the log mean of them.

Different expressions are reported in the literature to evaluate the effective radial thermal conductivity λ_{er} [5] and the heat transport coefficient in the first layer near the tube wall h_W [6–9] for the pseudo-homogeneous fluid/solid phase.

For ideal models, momentum balance is expressed as:

$$\frac{dP}{dz} = \frac{f \cdot G \cdot \mu_g}{\rho_g d_p^2} \cdot \frac{(1-\varepsilon)^2}{\varepsilon^3} \tag{4.5}$$

where P is the reaction pressure, f the friction factor, G the superficial mass flow velocity, μ_g the gas mixture viscosity, d_p the equivalent catalyst particle diameter, and ε the void fraction of the packed bed.

Equations 4.2, 4.3, and 4.5, with boundary conditions in the inlet section, are the ODE set to be solved to calculate concentrations, temperature and pressure profiles along the reactor axial coordinate.

Obviously, the ideal model is very simple and consequently solutions obtained are approximate. Usually, this type of model is used only when a "first tentative" solutions is required.

If a more accurate analysis of the phenomena inside the packed bed has to be performed, some of ideal model assumptions has to be released.

First of all, the axial gradients of concentrations and temperature lead to back-mixing effects and diffusive contributions to mass and heat transport, which bring about axial gradients smoother than those obtained by an ideal model, have to be included.

Taken as reference the mass balance, the Eq. 4.2 is modified as follows:

$$\varepsilon D_{ea} \frac{d^2 C_A}{dz^2} - u_s \frac{dC_A}{dz} = \rho_B r_A \tag{4.6}$$

where D_{ea} is the effective axial diffusivity, usually calculated in terms of the axial Pe number based on pellets diameter:

$$Pe_a = \frac{u_i \cdot d_p}{D_{ea}} \tag{4.7}$$

where u_i is the interstitial velocity of the gas mixture. Pe_a value lies within the range 0.5–2 [1].

In this kind of models second-order differential terms appear in mass, energy, and momentum balances and two boundary conditions on the axial coordinate have to be imposed. The set of differential equations together with the boundary conditions represents a boundary value problem (BVP), since the constraints are imposed at the inlet and at the outlet of the tubular reactor.

If radial variations are considered as well, the hypothesis of plug flow and consequently the one-dimensional nature of model equations falls through. Two-dimensional models are usually required for energy balances in packed bed reactors, since the scarce conductivity of packed bed leads to strong temperature radial changes.

As an example, the mass and energy balances for a two-dimensional pseudo-homogeneous packed bed reactor model are represented by the following equations:

$$\varepsilon D_{\mathrm{er}}\left(\frac{\partial^2 C_{\mathrm{A}}}{\partial r^2}+\frac{1}{r}\frac{\partial C_{\mathrm{A}}}{\partial r}\right)-u_{\mathrm{s}}\frac{\mathrm{d}C_{\mathrm{A}}}{\mathrm{d}z}=\rho_{\mathrm{B}}r_{\mathrm{A}} \tag{4.8}$$

$$-\lambda_{\mathrm{er}}\left(\frac{\partial^2 T}{\partial r^2}+\frac{1}{r}\frac{\partial T}{\partial r}\right)+u_{\mathrm{s}}\rho_{\mathrm{g}}c_{\mathrm{p}}\frac{\mathrm{d}T}{\mathrm{d}z}=(-\Delta H)\rho_{\mathrm{B}}r_{\mathrm{A}} \tag{4.9}$$

where axial mixing is neglected and radial convective term is negligible with respect to diffusive one. D_{er} and λ_{er} are the effective radial mass diffusivity [10] and thermal conductivity, respectively.

The set to be solved is composed by partial differential equations (PDEs), which can be handled by different approaches, as the Finite Elements Method or the Orthogonal Collocations.

4.1.1.2 Heterogeneous Models

For heterogeneous models, gas and solid phases are considered separately and mass, energy, and momentum balances have to be imposed for both phases. This approach has to be followed when:

- reactions are quick on the catalyst pellets surface, leading to large temperature and concentration gradients between the gas bulk and the solid surface;
- thermal conductivities of gas and solid phases are very different.

If the system to be modeled verifies one of these conditions at least, the heterogeneous approach has to be followed since a pseudo-homogeneous model would lead to unacceptable errors.

The simplest heterogeneous model considers only the interfacial gradients between solid and fluid phases, imposing mass and energy balances for all phases involved and introducing mass and heat transport from the gas bulk to catalyst surface and vice versa. Typical mass and energy balances equations are:

Fluid Phase:

$$-u_{\mathrm{s}}\frac{\mathrm{d}C_{\mathrm{A}}}{\mathrm{d}z}=K_{\mathrm{g}}a\left(C_{\mathrm{A}}-C_{\mathrm{A}}^{\mathrm{S}}\right) \tag{4.10}$$

$$u_{\mathrm{s}}\rho_{\mathrm{g}}c_{\mathrm{p}}\frac{\mathrm{d}T}{\mathrm{d}z}=h_{\mathrm{f}}a\left(T^{\mathrm{S}}-T\right)-4\frac{U}{d_{\mathrm{t}}}(T-T_{\mathrm{r}}) \tag{4.11}$$

Solid Phase:

$$\rho_{\mathrm{B}}r_{\mathrm{A}}=K_{\mathrm{g}}a\left(C_{\mathrm{A}}-C_{\mathrm{A}}^{\mathrm{S}}\right) \tag{4.12}$$

$$(-\Delta H)\rho_B r_A = h_f a(T^S - T) \tag{4.13}$$

where C_A^S and T^S are A-component concentration and temperature on catalyst surface and K_g and h_f are the mass and heat transport coefficient between gas and solid phase and a is the ratio between the area of the total external surface of particles and the reactor volume. Reaction occurs on the catalyst surface and reaction terms in the solid phase mass/energy balances, Eqs. 4.12, 4.13, have to be calculated at the temperature and components concentrations on the particle surface. The assumption made is that the catalyst is not-porous and fluid components cannot diffuse through solid material.

However, usually catalysts are porous materials to strongly increase the active reaction area per unit of reactor volume. Therefore, the intra-particle gradients, i.e., the mass and heat transfer inside the catalyst particle, have to be considered as well. Fluid phase equations are the same as Eqs. 4.10, 4.11, while solid phase equations have to be modified as follows:

Solid Phase:

$$\frac{D_e}{\xi^2}\frac{d}{d\xi}\left(\xi^2\frac{dC_{A,s}}{d\xi}\right) - \rho_s r_A(C_{A,s}, T_s) = 0 \tag{4.14}$$

$$\frac{\lambda_e}{\xi^2}\frac{d}{d\xi}\left(\xi^2\frac{dT_s}{d\xi}\right) + \rho_s(-\Delta H) r_A(C_{A,s}, T_s) = 0 \tag{4.15}$$

where $C_{A,s}$ and T_s are the concentration and temperature at the radial coordinate ξ inside the solid particle, assumed to be spherical, directed from the center to the external surface.

The terms D_e and λ_e are the effective diffusivity and thermal conductivity in the solid particle: methods to calculate these coefficients are reported in the literature [1, 11].

A different approach to determine the intra-particle effect is based on the effectiveness factor defined as:

$$\eta = \frac{\frac{1}{V_p}\int r_A(C_{A,s}, T_s)\,dV_p}{r_A(C_A^S, T^S)} \tag{4.16}$$

where V_p is the pellet volume. Effectiveness factor is the ratio between the real rate of reaction observed in the presence of pore diffusion resistance and the rate of reaction that should be observed if the whole particle be at surface conditions. Its value is within the range (0, 1) and it multiplies the rate of reaction at surface conditions to consider the reduction of reaction rate due to the diffusion resistance. Solid phase mass/energy, Eqs. 4.14 and 4.15, can be simplified as follows:

$$K_g a\left(C_A - C_{A,s}^S\right) = \eta\rho_B r_A\left(C_{A,s}^S, T_s^S\right) \tag{4.17}$$

$$h_f a\left(T_s^S - T\right) = \eta \rho_B(-\Delta H) r_A\left(C_{A,s}^S, T_s^S\right) \tag{4.18}$$

The most comprehensive packed bed reactor model is formulated accounting of:

- interfacial gradient between solid and fluid phases;
- intra-particle gradients inside catalyst pellets;
- radial mixing, leading to a 2D formulation;
- axial mixing, even if such an effect could be usually neglected.

The equations of the 2D heterogeneous model, neglecting axial mixing, are the following ones:

$$u_s \frac{\partial C_A}{\partial z} = \varepsilon D_{er}\left(\frac{\partial^2 C_A}{\partial r^2} + \frac{1}{r}\frac{\partial C_A}{\partial r}\right) - K_g a\left(C_A - C_{A,s}^S\right) \tag{4.19}$$

$$u_s \rho_g c_p \frac{\partial T}{\partial z} = \lambda_{er,f}\left(\frac{\partial^2 T}{\partial r^2} + \frac{1}{r}\frac{\partial T}{\partial r}\right) + h_f a\left(T_s^S - T\right) \tag{4.20}$$

$$K_g a\left(C_A - C_{A,s}^S\right) = \eta \rho_B r_A\left(C_{A,s}^S, T_s^S\right) \tag{4.21}$$

$$h_f a\left(T_s^S - T\right) = \eta \rho_B(-\Delta H) r_A\left(C_{A,s}^S, T_s^S\right) + \lambda_{er,s}\left(\frac{\partial^2 T_s}{\partial r^2} + \frac{1}{r}\frac{\partial T_s}{\partial r}\right) \tag{4.22}$$

In order to solve the PDE set, boundary conditions on axial and radial coordinates are required.

Obviously, such a complex formulation is developed only if an extremely rigorous analysis is required.

4.1.2 Pd-based Hydrogen Selective Membrane Modeling

4.1.2.1 Pd-based Membrane Permeability Definition

In modeling membrane reactors, a proper definition of membrane performance in terms of hydrogen flux is crucial. In fact, hydrogen permeation through selective membranes is a leading phenomenon for membrane reactors, and errors in assessing membrane behavior would lead to unreliable results.

In Chap. 2, H_2 permeation mechanisms through selective membranes are extensively reported. In this paragraph, the strategies to develop permeation models for Pd-based membranes and experimental protocol to evaluate parameters appearing in model equations are described.

As for the solution-diffusion mechanism, the hydrogen permeation through a palladium layer is a complex process consisting of adsorption and dissociation of hydrogen molecules, followed by diffusion of hydrogen atoms through the metal

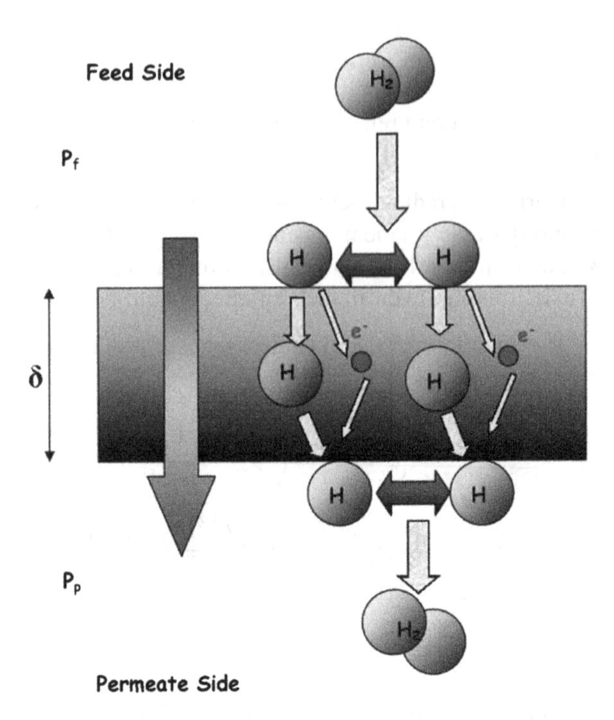

Fig. 4.4 Hydrogen permeation mechanism through Pd layer

lattice, recombination of hydrogen atoms at the low pressure side and desorption of molecular hydrogen, according to the following 6-step activated mechanism (see Fig. 4.4) [12]:

1. adsorption of hydrogen molecules on the high pressure side of the membrane surface;
2. dissociation of the chemisorbed molecules into atomic $H^\bullet$ species, consisting of a proton and an electron;
3. dissolution of the hydrogen atoms $H^\bullet$ (proton and electron) into the lattice of the metal;
4. diffusion of the hydrogen atoms $H^\bullet$ (proton and electron) through the lattice from the high hydrogen pressure of the membrane side to the low hydrogen pressure of the membrane surface;
5. re-combination of protons and electrons and re-association of two atomic $H^\bullet$ species with formation of hydrogen molecules H_2 at the low hydrogen pressure membrane surface;
6. desorption of hydrogen molecules from the low hydrogen pressure membrane side to the bulk.

Depending on the thickness of the membrane, step 5 (diffusion of hydrogen atoms through the lattice) results the controlling step in many cases. The diffusion of hydrogen atoms through the lattice is ruled by the Fick's law:

$$J = D \cdot \frac{dc_H}{dx} \tag{4.23}$$

where J is the hydrogen flux through the lattice, D is the hydrogen diffusion coefficient, and c_H is the hydrogen concentration.

For thick membranes (thickness of tens μm), the dissociative chemisorption of hydrogen on the membrane surface can be considered very fast compared with diffusion of atomic hydrogen in the membrane. Thermodynamic equilibrium conditions between hydrogen atoms dissolved at the surface of the membrane and hydrogen in the gas phase can be assumed and the following relation between the concentration of hydrogen atoms in Pd and the hydrogen partial pressure in the gas phase is easily get:

$$c_H = S \cdot p_{H_2}^{0.5} \tag{4.24}$$

where S is the hydrogen solubility and p_{H_2} the hydrogen partial pressure. The exponent 0.5 accounts for the dissociation of the hydrogen molecules into hydrogen atoms.

From Eqs. 4.23 and 4.24, the following equation can be derived for the hydrogen flux at constant temperature through a dense palladium membrane:

$$J_{H_2} = \frac{B_H}{\delta} \cdot \left(p_{H_2,r}^n - p_{H_2,p}^n\right) \quad n = 0.5 \tag{4.25}$$

where J_{H_2} is the hydrogen flux through the membrane, B_H the hydrogen permeability, δ the membrane thickness, $p_{H_2,r}$ and $p_{H_2,p}$ the hydrogen partial pressures in the retentate side (high hydrogen partial pressure) and in the permeate side (low hydrogen partial pressure), respectively.

Experimental values of n have been determined in the range 0.5–1 [13]. When $n = 0.5$, the hydrogen flux follows the so-called Sieverts' law which applies only where the hydrogen to metal atomic ratio is quite small (H/Pd $\ll$ 1).

As for the influence of temperature on hydrogen permeability, the relationship between the hydrogen permeation rate and the temperature can be described by the Arrhenius law:

$$B_H - P_0 \cdot \exp\left(-\frac{E_a}{R \cdot T}\right) \tag{4.26}$$

where P_0, E_a, R, and T are the pre-exponential factor, the apparent activation energy, the universal gas constant, and the absolute temperature, respectively.

From Eqs. 4.25 and 4.26, we get the following expression to describe the H_2 flux through membrane in terms of membrane type, thickness and H_2 partial pressure driving force:

$$J_{H_2} = \frac{P_0 \cdot \exp\left(-\frac{E_a}{R \cdot T}\right)}{\delta} \cdot \left(p_{H_2,r}^n - p_{H_2,p}^n\right) \tag{4.27}$$

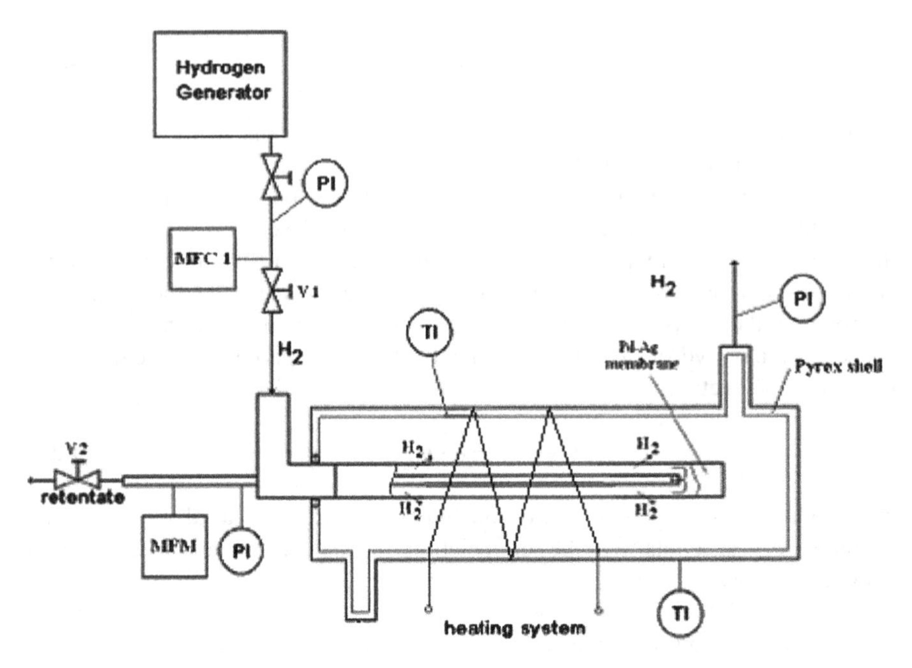

Fig. 4.5 Scheme of membrane testing experimental apparatus

4.1.2.2 Membrane Permeability Experimental Apparatus and Procedure

In Eq. 4.27 there are two parameters to be evaluated by an experimental procedure: P_0 and E_a.

An experimental apparatus for membrane permeability tests is shown in Fig. 4.5.

It is composed by:

- A mass flow controller MFC1 to regulate the hydrogen feed;
- A flow meter in the outlet permeation side to measure the permeated hydrogen stream;
- Three pressure gages (PI) for measuring the pressure at inlet and outlet of the membrane tube and at shell outlet (two pressure gages are assembled in the lumen since the system is suitable for the membrane reactor testing as well, for which a pressure drop evaluation is necessary);
- A shut-off valve (V1) on the hydrogen generator line while the throttle valve (V2) on the retentate line allows adjusting the pressure inside the membrane.
- A remote-controlled heating system consisting of a platinum coil wire wrapped around the membrane tube for assuring isothermal conditions. Thermocouples (TI) monitor the temperature.

A permeator module, developed by Angelo Basile's research group of ITM-CNR, is schematized in Fig. 4.6.

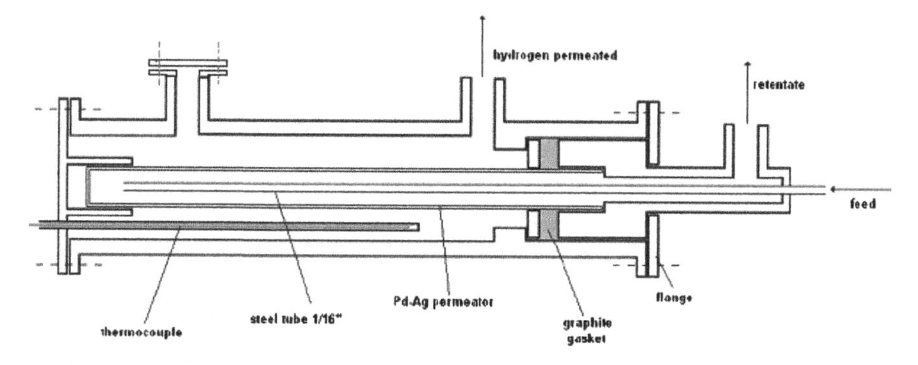

Fig. 4.6 Scheme of the permeator module developed by ITM-CNR

The permeation tests are carried out by the following procedure:

1. the temperature is regulated by the remote-controlled heating system;
2. when the temperature of the system is stable, a hydrogen stream is sent into the retentate side through the mass flow controllers MFC1;
3. then, the pressure of the lumen (retentate side) is regulated by the throttle valve (V2), while the pressure in the shell can be maintained at a fixed pressure;
4. when the operating conditions are stable, the hydrogen flow permeated through the membrane is measured by the flow meter.

Usually, the task of experimental phase is threefold:

1. verification of the permeation mechanism according to the Sieverts' law;
2. verification of the Arrhenius type temperature dependence of the membrane permeability;
3. evaluation of the permeability expression parameters.

About point 1, the scope is to verify if exponent n in Eq. 4.25 is equal to 0.5. The following testing procedure has to be followed:

1. at a fixed temperature, the permeated hydrogen flux for various pressure differences between shell and lumen is measured;
2. the data collected are reported as J_{H_2} vs. $\left(p_{H_2,r}^{0.5} - p_{H_2,p}^{0.5}\right)$ function;
3. if the data arrange on a line, the assumption of the Sieverts' law validity is verified.

The test should be repeated for three temperatures at least, in order to verify Sieverts' law within an operating temperature range.

Then, the Arrhenius type temperature dependence of membrane permeability has to be verified as well, and at the same time the values of parameters P_0 and E_a have to be assessed. The experimental procedure is:

1. at a fixed pressure difference between shell and lumen, the hydrogen permeated flux is measured for different operating temperature.

2. Assuming the Arrhenius permeability dependence on the temperature and the Sieverts' law, the hydrogen flux can be described by:

$$J_{H_2} = \frac{P_0 \cdot \exp\left(-\frac{E_a}{R \cdot T}\right)}{\delta} \cdot \left(p_{H_2,r}^{0.5} - p_{H_2,p}^{0.5}\right) \quad (4.28)$$

Considering that pressure driving force and membrane thickness are constant, Eq. 4.28 can be written as:

$$J_{H_2} = A \cdot P_0 \exp\left(-\frac{E_a}{RT}\right) \quad (4.29)$$

where A is a known constant. The Eq. 4.29 has to be linearized for the data fitting, therefore:

$$-\ln(J_{H_2}) = B + \frac{C}{T} \quad (4.30)$$

where:

$$B = -\ln(A \cdot P_0)\ C = \frac{E_a}{R} \quad (4.31)$$

3. The data collected are reported as $-\ln(J_{H_2})$ vs. $\frac{1}{T}$. If the data arrange on a straight line, the Arrhenius dependence on the temperature is verified and the values of P_0 and E_a can be derived from the slope and the intercept of the fitting straight line. Refer to Sect. 10.5.2 as an example of application for this procedure.

At the end of the experimental procedure, a validated selective hydrogen membrane permeability is available for membrane reactors modeling.

4.2 Integrated Membrane Reactor Models

In this section, strategies and equations used in membrane reactors modeling are presented. A detailed model applied to natural gas steam reforming is reported in Chap. 5.

An integrated membrane reactor (IMR) is usually composed by a tube, packed with catalyst pellets, inside which one or more membrane modules are assembled (Fig. 4.7). The catalyst could be also packed in the inner tube, but generally it is preferred assembling the membrane in such a way that the higher pressure (reaction pressure) is imposed outside membrane wall.

In order to study the reactor behavior, mass, energy, and momentum balances have to be imposed for both reaction zone, where catalyst is packed and reactions are promoted, and permeation zone, where an inert gas is sent to sweep the hydrogen permeated through the selective membranes.

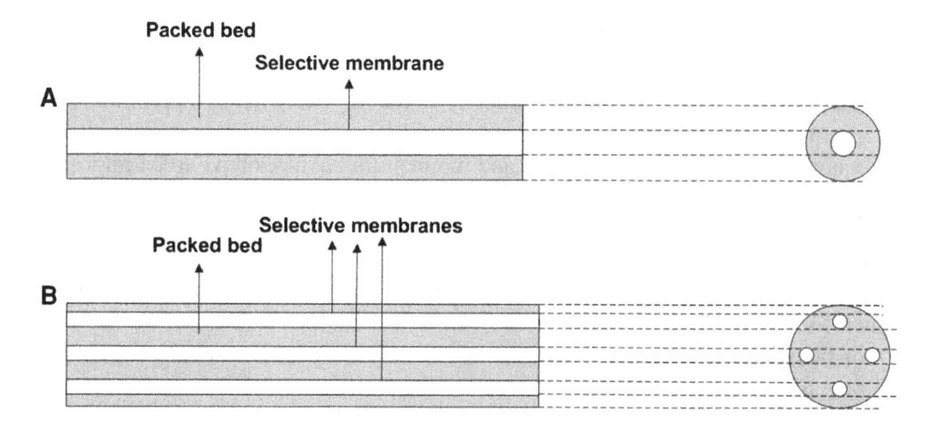

Fig. 4.7 One membrane module (**a**) and multiple membrane (**b**) reactor configuration

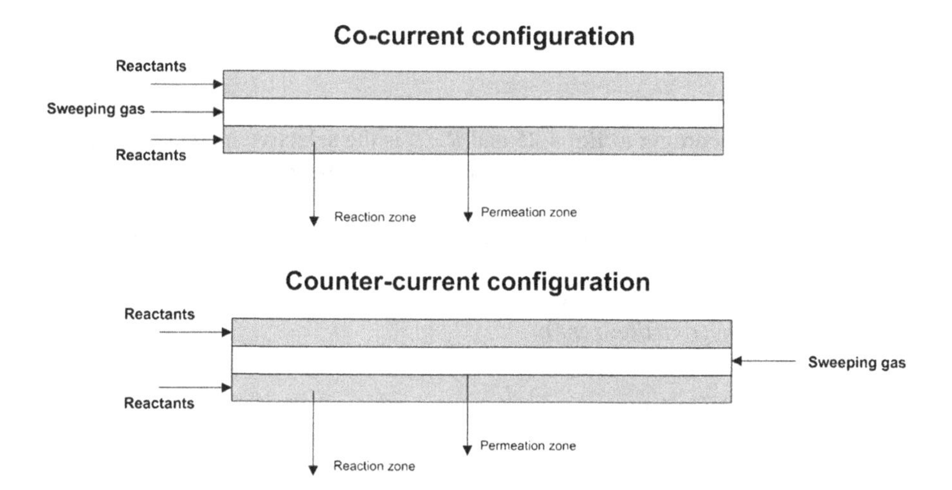

Fig. 4.8 Co-current and counter-current configurations

Two configurations are possible (Fig. 4.8):

1. co-current configuration, where reactants in the reaction zone and sweeping gas in the permeation zone flow in the same direction;
2. counter-current configuration, where reactants and sweeping gas flow in opposite directions.

A potential alternative configuration is the application of a vacuum pump in the inner zone to reduce membrane downstream pressure, supporting hydrogen permeation without feeding a sweep gas.

In the following sections both one- and two-dimensional models are described.

4.2.1 One-Dimensional Model

For sake of simplicity, an ideal model is taken as example to approximately describe the IMR behavior.

As aforementioned, mass, energy, and momentum balances have to be imposed both in reaction and in permeation zone.

Reaction rate equation and all physical properties as heats of reactions, thermal transport coefficient, density of packed bed and fluid mixture, fluid mixture viscosity, reactor and catalyst particles diameters, void fraction, etc., have to be known for the specific system.

For the reaction zone, the equations of the model are the following ones:

Mass Balance (reaction zone):

$$\frac{\mathrm{d}(u_s c_i)}{\mathrm{d}z} = \rho_B r_i \quad \text{for } i \neq \mathrm{H_2} \tag{4.32}$$

$$\frac{\mathrm{d}(u_s c_{\mathrm{H_2}})}{\mathrm{d}z} = \rho_B r_{\mathrm{H_2}} - \frac{2}{R_{\mathrm{mem}}} J_{\mathrm{H_2}} \quad \text{for } i = \mathrm{H_2} \tag{4.33}$$

where r_i and r_{H_2} are reaction rates for i-component and for hydrogen, respectively. $J_{\mathrm{H_2}}$ is expressed according to Eq. 4.25 and R_{mem} is the selective membrane radius. The term u_s is included in the derivative since in IMR the gas velocity can change both for gas volume variation due to reactions and for hydrogen flow outgoing from the reaction environment through the selective membrane. Obviously, hydrogen mass balance in reaction zone must include the hydrogen flux leaving the reaction environment thanks to selective membrane integration.

Energy Balance (reaction zone):

$$u_s \rho_g c_p \frac{\mathrm{d}T_{\mathrm{reac}}}{\mathrm{d}z} = \sum_{i=1}^{n_r} (-\Delta H_i) \rho_B r_i - 4\frac{U}{d_t}(T_{\mathrm{reac}} - T_r) - 4\frac{d_{\mathrm{mem}} \cdot U_1}{d_t^2}(T_{\mathrm{reac}} - T_{\mathrm{perm}}) \tag{4.34}$$

where T_{reac} and T_{perm} are reaction and permeation zone temperature, respectively, d_{mem} is the membrane diameter, n_r is the number of reactions involved, and U_1 is the overall heat transfer coefficient between reaction and permeation zone, expressed as follows:

$$U_1 = \left[\frac{1}{h_W} + \frac{\delta}{\alpha_{\mathrm{mem}}} + \frac{d_{\mathrm{mem,o}}}{d_{\mathrm{mem},i}} \cdot \frac{1}{h_{\mathrm{perm}}}\right]^{-1} \tag{4.35}$$

In Eq. 4.35 α_{mem} is the membrane thermal conductivity, $d_{\mathrm{mem},i}$ and $d_{\mathrm{mem,o}}$ are the internal and external membrane diameter, and h_{perm} is the convective heat transport coefficient in the permeation zone.

In Eq. 4.34, the effect of enthalpy flux associated to hydrogen outgoing from the reaction zone through the membranes is neglected.

Momentum Balance (reaction zone): In reaction zone, the momentum balance can be expressed according to Eq. 4.5.

As for the permeation zone, balances are represented by following equations:

Mass Balance (permeation zone):

$$\frac{\mathrm{d}\left(u_{s,\mathrm{perm}} c_{i,\mathrm{perm}}\right)}{\mathrm{d}z} = 0 \quad \text{for } i \neq \mathrm{H_2} \tag{4.36}$$

$$\pm\frac{\mathrm{d}\left(u_{s,\mathrm{perm}} c_{\mathrm{H_2},\mathrm{perm}}\right)}{\mathrm{d}z} = \frac{2}{R_{\mathrm{mem}}} J_{\mathrm{H_2}} \quad \text{for } i = \mathrm{H_2} \tag{4.37}$$

where $u_{s,\mathrm{perm}}$ is the superficial gas velocity calculated at permeation zone operating conditions. The sign $\pm$ refers to co-current (+) or counter-current ($-$) mode.

Energy Balance (permeation zone):

$$\pm u_{s,\mathrm{perm}} \cdot \rho_{g,\mathrm{perm}} \cdot c_{p,\mathrm{perm}} \frac{\mathrm{d}T_{\mathrm{perm}}}{\mathrm{d}z} = 4\frac{U_1}{d_{\mathrm{mem}}}\left(T_{\mathrm{reac}} - T_{\mathrm{perm}}\right) \tag{4.38}$$

where $\rho_{g,\mathrm{perm}}$ and $c_{p,\mathrm{perm}}$ are gas density and specific heat calculated at permeation zone operating conditions, respectively. Likewise to reaction zone energy balance, the enthalpy flow leaving the reaction zone with the permeated hydrogen flux is neglected.

Momentum Balance (permeation zone): Usually, pressure drops in permeation zone can be neglected. However, typical expressions are available for fluids flowing in a tube, both in laminar and turbulent conditions, and could be applied.

The ODE set composed by Eqs. 4.32–4.34, 4.5, 4.36–4.38 can be solved by a numerical procedure.

The following boundary conditions can be imposed:

$$z = 0: \quad \begin{aligned} & u_s c_i = \left(u_s c_i\right)_{\mathrm{in}} \\ & T_{\mathrm{reac}} = T_{\mathrm{reac,in}} \\ & P = P_{\mathrm{in}} \\ & u_{s,\mathrm{perm}} c_{\mathrm{H_2},\mathrm{perm}} = 0 \quad \text{Co-current configuration} \\ & T_{\mathrm{perm}} = T_{\mathrm{perm,in}} \quad \text{Co-current configuration} \end{aligned} \tag{4.39}$$

$$z = L: \quad \begin{aligned} & u_{s,\mathrm{perm}} c_{\mathrm{H_2},\mathrm{perm}} = 0 \quad \text{Counter-current configuration} \\ & T_{\mathrm{perm}} = T_{\mathrm{perm,in}} \quad \text{Counter-current configuration} \end{aligned} \tag{4.40}$$

In the case of co-current configuration, ODE set is solved by a numerical method with boundary conditions at inlet section ($z = 0$); if a counter-current design has to be simulated, the problem is a BVP since boundary conditions for reaction zone are imposed at inlet section ($z = 0$) while for permeation zone they are imposed at outlet section ($z = L$). Different methods can be applied to solve BVP; the most used one is the "shooting method" by which:

1. permeation zone inlet conditions ($z = 0$) are assumed;
2. ODE set is solved;
3. if the real boundary conditions (4.40) are verified, the inlet boundary conditions assumed at point 1 are the final results, otherwise other boundary conditions have to be imposed at point 1.

4.2.2 Two-Dimensional Model

Usually, IMR are characterized by marked compositions and temperature radial profiles, which affect the reactor behavior. Therefore, the radial mixing effect should be always considered in IMR modeling, and 2D model are certainly more reliable and realistic.

A 2D pseudo-homogeneous model for IMR is presented here.

Usually, the radial diffusive transport term is taken into account for mass and energy balances, while for momentum balance it could be neglected since the presence of catalyst pellets produces a mixing effect leading to a uniform gas velocity radial profile. Moreover, the radial diffusive terms are always much greater than the convective radial terms, which can be neglected.

For the reaction zone, mass and energy balances are Eqs. 4.8 and 4.9, with a gas velocity term u_s generally dependent on axial coordinate and included in the derivatives:

$$\frac{\varepsilon D_{\mathrm{er}}}{u_{\mathrm{s}}}\left(\frac{\partial^2 (u_{\mathrm{s}} c_i)}{\partial r^2} + \frac{1}{r}\frac{\partial (u_{\mathrm{s}} c_i)}{\partial r}\right) - \frac{\mathrm{d}(u_{\mathrm{s}} c_i)}{\mathrm{d}z} = \rho_{\mathrm{B}} r_i \tag{4.41}$$

$$-\lambda_{\mathrm{er}}\left(\frac{\partial^2 T_{\mathrm{reac}}}{\partial r^2} + \frac{1}{r}\frac{\partial T_{\mathrm{reac}}}{\partial r}\right) + u_{\mathrm{s}}\rho_{\mathrm{g}} c_{\mathrm{p}}\frac{\mathrm{d}T_{\mathrm{reac}}}{\mathrm{d}z} = \sum_{i=1}^{n_{\mathrm{r}}}(-\Delta H_i)\rho_{\mathrm{B}} r_i \tag{4.42}$$

In the permeation zone, radial mixing can be neglected and Eqs. 4.36–4.38 can be used.

Boundary conditions to solve the PDE set are:

$$z = 0,\ \forall r:$$

$$\begin{aligned}
& u_{\mathrm{s}} c_i = (u_{\mathrm{s}} c_i)_{\mathrm{in}} \\
& T_{\mathrm{reac}} = T_{\mathrm{reac,in}} \\
& P = P_{\mathrm{in}} \\
& u_{\mathrm{s,perm}} c_{\mathrm{H_2,perm}} = 0 \quad \text{Co-current configuration} \\
& T_{\mathrm{perm}} = T_{\mathrm{perm,in}} \quad \text{Co-current configuration}
\end{aligned} \tag{4.43}$$

$z = L, \forall r$

$$u_{s,perm} c_{H_2,perm} = 0 \quad \text{Counter-current configuration} \tag{4.44}$$
$$T_{perm} = T_{perm,in} \quad \text{Counter-current configuration}$$

$r = r_t, \forall z :$

$$\frac{\partial(u_s \cdot c_i)}{\partial r} = 0$$
$$\lambda_{er} \cdot \frac{\partial T_{reac}}{\partial r} = U \cdot \left(T_w - T_{reac|r_t}\right) \tag{4.45}$$

$r = r_{mem}, \forall z :$

$$\frac{\partial(u_s \cdot c_i)}{\partial r} = 0 \quad \text{for } i \neq H_2$$
$$\frac{d_p}{Pe_{mr}} \cdot \frac{\partial(u_s \cdot c_{H_2})}{\partial r} = J_{H_2} \tag{4.46}$$
$$\lambda_{er} \cdot \frac{\partial T_{reac}}{\partial r} = U_1 \cdot \left(T_{reac|r_{mem}} - T_{perm}\right)$$

where r_t and r_{mem} are the tube and membrane radii, T_W is the reactor tube wall temperature, Pe_{mr} is the mass effective radial Peclet number calculated by equation reported by Kulkarni [10] and valid for Reynolds number greater than 1000.

The PDE set can be solved by using different numerical approaches, as the Finite Elements Method or the Orthogonal Collocations.

By the 2D mathematical model, axial and radial profiles both of components concentrations and of reactor operating temperature can be calculated.

4.3 Modeling of Staged Membrane Reactor

Integrating selective membrane in a critical environment as a packed bed tubular reactor could lead to a series of design, operating and maintenance problems. Therefore, an alternative configuration is analyzed in the present section. Hydrogen selective membrane is assembled in separation modules located downstream to reaction units. This configuration, called *staged membrane reactor* (SMR) and shown in Fig. 4.9, is composed by a series of reaction-separation units: the reactant feedstock is fed to the first reactor where it is partially converted into hydrogen; then hydrogen is recovered through a selective membrane separation module, while the retentate is sent to the next step or recycled to the first module. Of course, it is possible to replicate the SMR until the desired reactant conversion is achieved.

SMR configuration is more deeply modeled and assessed in Chap. 5 for natural gas steam reforming. Here, equations and boundary conditions of a model useful to describe plant behavior are presented.

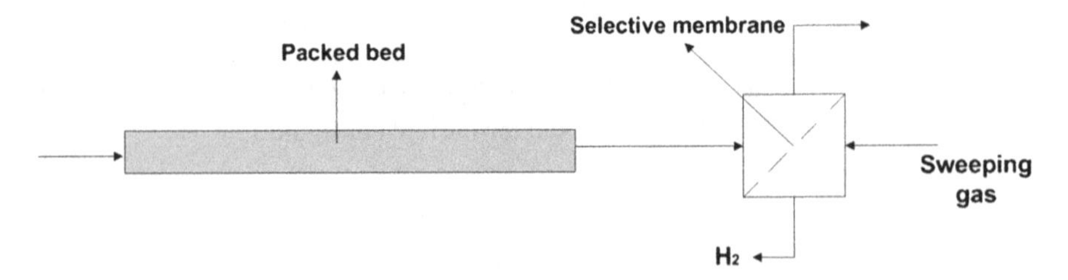

Fig. 4.9 Staged membrane reactor (SMR) draft

Both reactor and separation module have to be separately modeled, then they are connected by boundary conditions.

The reactor model equations are those reported in Sect. 4.1.1, since a traditional packed bed reactor has to be simulated.

Then, the stream leaving the reactor is fed to the separation module, which can be modeled by the following equations (only mass balances are used):

High pressure zone, where reactor output stream is fed:

$$\frac{\mathrm{d}\left(u_{\mathrm{s,high}}c_{i,\mathrm{high}}\right)}{\mathrm{d}z_{\mathrm{perm}}} = 0 \quad \text{for } i \neq \mathrm{H_2} \tag{4.47}$$

$$-\frac{\mathrm{d}\left(u_{\mathrm{s,high}}c_{\mathrm{H_2,high}}\right)}{\mathrm{d}z_{\mathrm{perm}}} = \frac{2}{R_{\mathrm{mem}}}J_{\mathrm{H_2}} \quad \text{for } i = \mathrm{H_2} \tag{4.48}$$

Low pressure zone, where an inert gas to sweep out hydrogen permeated is fed:

$$\pm\frac{\mathrm{d}\left(u_{\mathrm{s,low}}c_{\mathrm{H_2,low}}\right)}{\mathrm{d}z_{\mathrm{perm}}} = \frac{2}{R_{\mathrm{mem}}}J_{\mathrm{H_2}} \tag{4.49}$$

where z_{perm} is the axial coordinate of the separation module, $u_{\mathrm{s,high}}$ and $u_{\mathrm{s,low}}$ are superficial gas velocity in high pressure and low pressure zone, $c_{\mathrm{H_2,high}}$ and $c_{\mathrm{H_2,low}}$ are hydrogen concentrations in high pressure and low pressure mixture streams. The sign $\pm$ in low pressure zone mass balance depends on co-current or counter-current configuration.

Boundary conditions are defined as follows:

- For the first reactor, boundary conditions are plant feedstock composition, temperature and pressure, while for the other reactors of the series they are the outcome conditions of the high pressure zone of upstream separator.
- For the separators, boundary conditions are:

$$z = 0$$

$$\begin{aligned} u_{\mathrm{s,high}}c_{\mathrm{H_2,high}} &= u_{\mathrm{s,reactor}}c_{\mathrm{H_2,reactor}} \\ u_{\mathrm{s,low}}c_{\mathrm{H_2,low}} &= 0 \quad \text{Co-current configuration} \end{aligned} \tag{4.50}$$

$$z = L$$
$$u_{s,low} c_{H_2,low} = 0 \quad \text{Counter-current configuration} \tag{4.51}$$

4.4 Conclusions

Reactor mathematical modeling is a powerful tool for simulating the physical and chemical phenomena occurring in a reactor. The improvement of hardware and software allows reactor designers to work with more and more detailed and reliable algorithms, by which it is possible to understand behavior and performance even before fabricating the reactor itself.

In the present chapter, membrane reactors modeling strategies have been presented, and various models of tubular fixed bed reactors, Pd-based membrane separators, IMR and SMR are reported and explained.

Different complexity levels can be applied, from ideal to much complex structured models accounting for radial and axial mixing, intra-particle components diffusion through catalyst particles, inter-facial gradients between solid and fluid phases. Of course, increasing the model complexity leads to more accurate solutions but a growing computational effort is required: understanding which assumptions can be made in each case without significantly affecting results is one of the most important designers' skill.

For membrane reactors, many crucial phenomena have to be included, as membrane permeability mechanism and hydrogen flux, reactions kinetics, heat and mass transport inside the reactor and from the external to the reactor. Therefore, a proper simulation certainly requires a deep study and a careful evaluation and definition of the system. The model developers have to work in a strict connection with test drivers, since reliable model parameters and coefficients definition is crucial: designers can address reactors experimentation clarifying which information from test-benches are required. At the same time, a proper model development allows the number of experimental tests to be reduced drastically to those ones required for a complete reactor validation.

By this chapter, the author does not pretend to run out a so wide topic as reactor modeling and simulation but he wants only to give to readers the basis for a further and deeper study.

References

1. Froment GF, Bischoff KB (1990) Chemical reactor analysis and design. Wiley, New York
2. Fogler HS (1999) Elements of chemical reactor engineering, 3rd edn. Prentice Hall, Upper Saddle River

3. De Falco M, Marrelli L, Basile A (2009) An industrial application of membrane reactors: modelling of the methane steam reforming reaction, Chapter 9. In: Simulation of membrane reactors. Nova Science Publishers Inc., New York, ISBN 978-1-60692-425-9
4. Tsotsas E, Schlünder E (1990) Heat transfer in packed beds with fluid flow: remarks on the meaning and the calculation of a heat transfer coefficient at the wall. Chem Eng Sci 45:819–837
5. De Wasch AP, Froment GF (1971) A two dimensional heterogeneous model for fixed bed catalytic reactors. Chem Eng Sci 26:629–634
6. Li C, Finlayson B (1977) Heat transfer in packed beds—a reevaluation. Chem Eng Sci 32:1055–1066
7. Yagi S, Wakao S (1959) Heat and mass transfer from wall to fluid in packed beds. AIChE J 5(1):79–85
8. Yagi S, Kunii D (1957) Studies on effective thermal conductivities in packed beds. AIChE J 3(3):373–380
9. Dixon A, Di Costanzo M, Soucy B (1984) Fluid-phase radial transport in packed beds of low tube-to-particle diameter ratio. Int J Heat Mass Transfer 27(10):1701–1713
10. Kulkarni BD, Doraiswamy LK (1980) Estimation of effective transport properties in packed bed reactors. Cat Rev Sci Eng 22(3):431–483
11. Satterfield CN (1970) Mass transfer in heterogeneous catalysis. M.I.T. Press, Cambridge
12. Lewis FA (1967) The palladium hydrogen system. Academic Press, London 94
13. Dittmeyer R, Höllein V, Daub K (2001) Membrane reactors for hydrogenation and dehydrogenation processes based on supported palladium. J Mol Catal A Chem 173:135–184

Chapter 5
Membrane Integration in Natural Gas Steam Reforming

Marcello De Falco

5.1 Process Description

The total hydrogen produced today (65 million of tonnes per year) is mainly used in:

- chemical industry, as reactant for the ammonia and methanol synthesis, for the hydrogenation of vegetable oil and as reductant to produce metals from their oxides;
- in the refinery processes as the hydrodesulfuration of sulfur compounds and hydrocracking processes. The deteriorating quality of crude oils, the more stringent petroleum product specifications, and environmental problems are leading to larger needs of hydrogen in hydroprocessing.

Moreover, there is an increasing interest about hydrogen as energy carrier and as feedstock for fuel cells, which further stimulates hydrogen industry.

Hydrogen can be obtained from different sources as fossil fuels (natural gas reforming, and coal gasification), renewable fuels (biomass), algae, and vegetables or water (electrolysis and thermo-chemical cycles). Many different energy sources can be used in most of these processes: heat from fossil fuel or nuclear reactors, electricity from several sources as solar energy.

However, hydrogen production through steam reforming of natural gas (NG) is undoubtedly the most applied process (48% of the total production), mainly for the lower cost.

Figure 5.1 reports hydrogen cost based on production processes. It is a worth assessment that:

- NG steam reforming produces the most cost-competitive hydrogen if the natural gas price is not too high (<5 \$/GJ).

M. De Falco (✉)
Faculty of Engineering, University Campus Bio-Medico of Rome,
Via Alvaro del Portillo 21, 00128 Rome, Italy
e-mail: m.defalco@unicampus.it

M. De Falco et al. (eds.), *Membrane Reactors for Hydrogen Production Processes*,
DOI: 10.1007/978-0-85729-151-6_5,

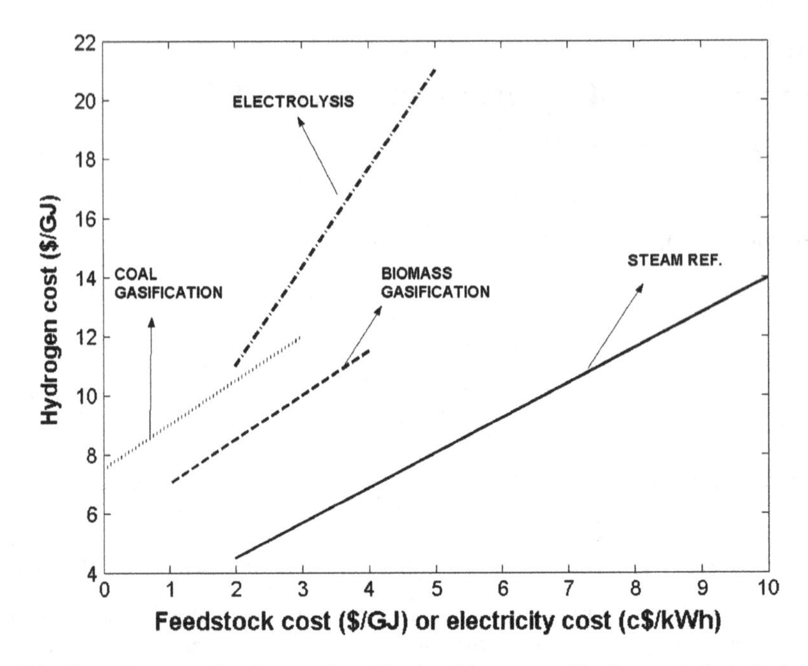

Fig. 5.1 Cost of steam reforming, coal gasification, biomass gasification, and electrolysis with respect to feedstock cost or electricity cost for electrolysis

- the cost of hydrogen produced by NG steam reforming is heavily dependent on the feedstock price, since the NG is both the reactant and the fuel of the process. For large plants, 70% of the production cost comes from the NG cost.
- obviously, a carbon tax will influence the cost of the hydrogen produced using NG and carbon, making more competitive biomass gasification and electrolysis.

Nowadays, NG steam reforming appears to be the only process able to produce large amounts of hydrogen at a competitive cost and to promote hydrogen technologies in the next years.

However, some uncertainty comes from the volatility of natural gas price (spot prices can double or triple in a short period of time) that affects the total cost of hydrogen production for a half.

Although the Department of Energy's Energy Information Administration forecasts natural gas prices to rise slowly till 2025 (see Fig. 5.2), the future is even less certain after 2025.

5.1.1 Reactions Thermodynamics

The main reactions involved in steam reforming process are:

$$CH_4 + H_2O \Leftrightarrow CO + 3H_2 \quad \Delta H_{298K} = 206.1\frac{kJ}{mol} \tag{5.1}$$

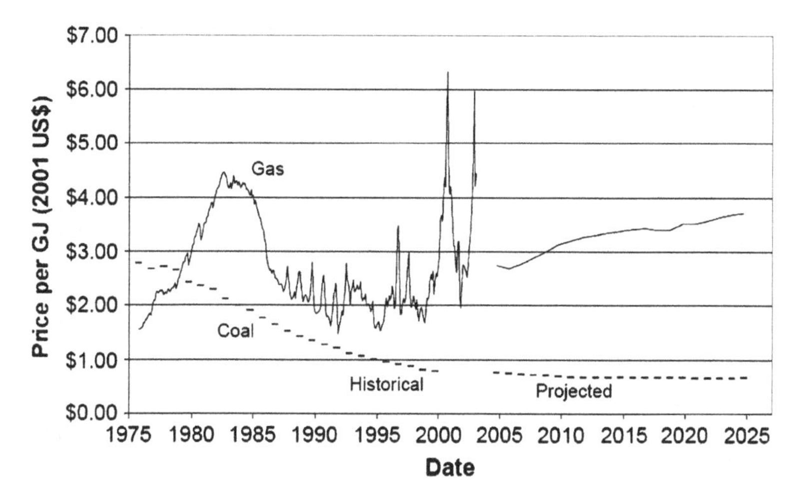

Fig. 5.2 Coal and NG price, historical and projected

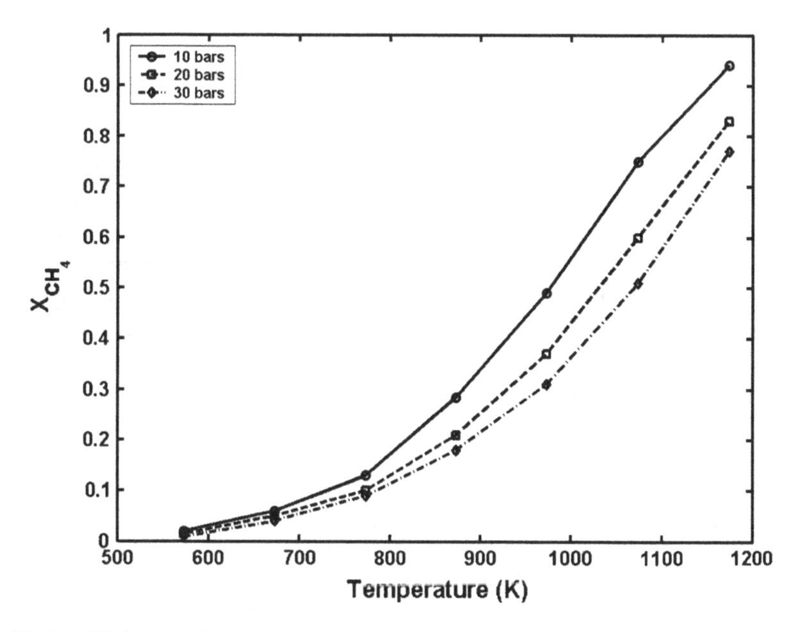

Fig. 5.3 Equilibrium methane conversion at S/C = 2

$$CO + H_2O \Leftrightarrow CO_2 + H_2 \quad \Delta H_{298K} = -41.15 \frac{kJ}{mol}. \tag{5.2}$$

The assumption of a natural gas composed by only methane is made. However, usually heavier hydrocarbon (ethane, propane, etc.) are cracked as the feedstock goes into the reactor tanks to the high process operating temperature (Fig. 5.3).

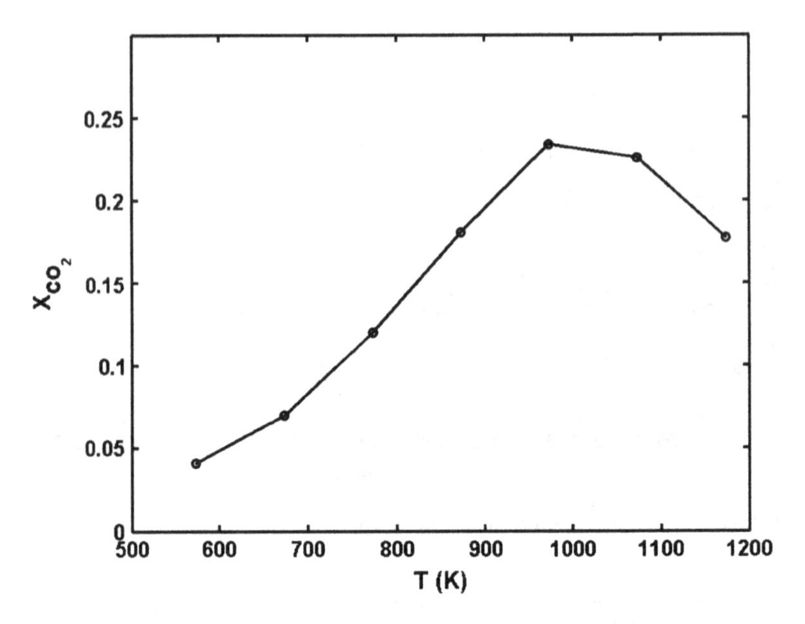

Fig. 5.4 Carbon dioxide yield versus temperature at S/C = 2

The first reaction is the steam reforming reaction (SR), which is an endothermic reaction thermodynamically promoted at high temperature and low pressure. The second reaction is known as water–gas shift reaction (WGS), an exothermic reaction favored at low temperature and not affected by the operating pressure.

Globally the steam reforming process should be supported by high temperature (in the steam reformer the task is the conversion of the methane, the WGS is supported by one or two steps water–gas shift reactor which follow the steam reformer, as described in the next paragraph) and by low pressure. However, usually the reactions are conducted at 25–40 bar in industrial plants to reduce the total volume of the devices and favor the heat transfer (Fig. 5.4).

Beyond the operating temperature and pressure, another important process variable is the steam to carbon ratio (S/C), defined as the ratio between the steam flow rate to the methane flow rate in the reactor feedstock (Fig. 5.5). On the basis of SR reaction stoichiometry S/C should be 2, but ratios greater than the stoichiometric value promotes the reactions and reduces the possibility of coke formation on the catalyst. In industry, the ratio is usually within the range 2.5–6 [1].

The effect of the temperature, pressure, and steam to carbon ratio on the thermodynamic equilibrium for the reactions 5.1 and 5.2 is reported in the following [2], where methane conversion X_{CH_4} and carbon dioxide yield X_{CO_2} are expressed according to:

$$X_{CH_4} = \frac{F_{CH_4,in} - F_{CH_4}}{F_{CH_4,in}} \tag{5.3}$$

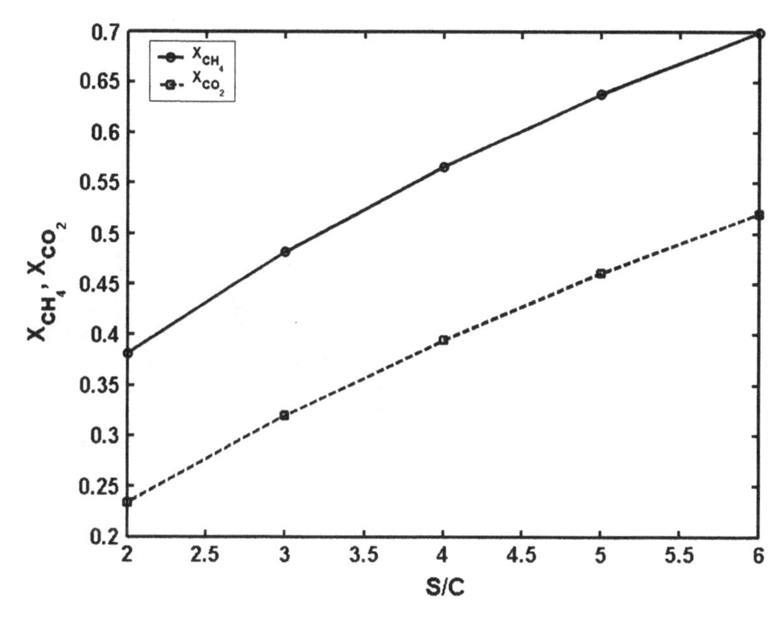

Fig. 5.5 Equilibrium conversions versus S/C at $T = 973$ K and $P = 20$ bar

$$X_{CO_2} = \frac{F_{CO_2} - F_{CO_2,in}}{F_{CH_4,in}}, \quad (5.4)$$

where $F_{CH_4,in}$ and $F_{CO_2,in}$ are inlet molar flow rates of methane and carbon dioxide, respectively.

5.1.2 Process Scheme

Because of the strong reactions endothermicity, steam reforming process is carried out by supplying a high heat flux. Two alternatives are available:

1. tubular fired reforming (see Fig. 5.6);
2. autothermal reforming.

In tubular fired reforming, process heat duty is supplied by burning a share of NG feedstock in burners placed at top, down, or sideways of a furnace. In autothermal reformers, a part of NG feedstock is directly burned in the first section of the reactor by adding an oxygen or enriched air stream and heat duty is supplied without any external source.

The autothermal process, described in Chap. 6, is competitive with tubular fired reforming only for large applications as GTL plants or if oxygen or enriched air are available at low cost since the use of natural air for methane combustion would not allow the high temperature required by reactions to be reached.

Fig. 5.6 NG tubular fired steam reforming plant (Courtesy from Tecnimont-KT)

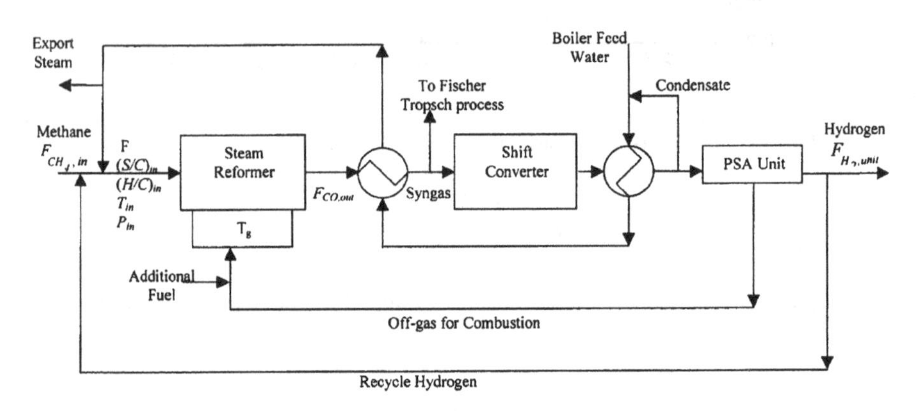

Fig. 5.7 Process flow diagram for the NG steam reforming

A typical flow-sheet of the tubular fired steam reforming process is shown in Fig. 5.7.

Natural gas is mixed with appropriate amounts of steam and recycled hydrogen before entering the reformer reactor. The recycle of an amount of H_2 produced is necessary to keep the catalyst in the early part of the reformer tubes in the reduced (active) state.

Reactions 5.1 and 5.2 occur in parallel in the steam reformer: the high temperature of the reactor, placed in a furnace, supports the steam reforming reaction to detriment of the WGS.

The hot syngas produced, composed by CH_4, H_2O, H_2, CO, and CO_2, is used to generate very high pressure (VHP) steam for mixing with the feed (internal use) and for "export" outside the unit. The cooled syngas is sent to a water gas shift converter: generally, the operating temperature is about 400–450°C to support the reaction kinetics over the iron oxides catalyst.

The H_2-rich exit stream from the shift converter is cooled to condensate the steam, producing the water recycled into the water boilers.

Then, the H_2 is separated from the off-gases (CH_4, CO, CO_2) in a pressure swing adsorption (PSA) section, which involves the adsorption of impurities onto a fixed bed of adsorbents at high pressure. The impurities are subsequently desorbed at low pressure into an off-gas stream. This operation allows an extremely pure hydrogen to be reached: H_2 purities over 99.999% can be achieved.

Off-gases, mixed with additional fuel, are sent to the burners to supply a part of the process heat duty.

If CO_2 sequestration is required for environmental reasons, a Methyl-Di-Ethanol-Amine (MDEA) unit has to be inserted to absorb the carbon dioxide and to separate it from the other off-gas mixture components.

The global efficiency of the traditional steam reformer plants, calculated as the ratio between the net heat value of the hydrogen stream produced and the total process heat duty requirement (reactor heat duty, steam generation, pre-heating of the reactant mixture, PSA, and MDEA) is typically within the range 65–85%, depending on the plant size.

5.1.3 Traditional Steam Reforming Process Drawbacks

The main steam reforming process drawbacks are:

- a significant fuel consumption to heat up the reactor. Methane unconverted in the reactor plus an additional methane stream have to be burnt to supply the heat required by reactions: the average heat flux between the oven and the catalyst packed bed is about 80 kW/m^2 for tubular fired reactors; therefore for each tube of the reforming reactor, 10 m long and with a diameter of 10 cm, the heat duty is about 250 kW. Considering that in a traditional industrial plant hundreds of parallel tubes are installed, the total heat duty is some tens of MWth. Moreover, the high NG fuel consumption is the main reason of the strong dependence of the final H_2 cost on the NG market price (Fig. 5.1) and of the consequent large uncertainty on the expected H_2 production cost.
- strong mechanical stresses on reformer tubes due to the high temperature with large thermal gradient in the axial direction. The high operating temperature forces to use expensive high alloy steels whereas the large thermal gradient in axial direction, together with the pressure (25–30 bar) in the tube, produce hard stresses on materials.
- the use of a furnace requires a good operational experience.
- the scarce catalyst effectiveness, mainly due to the small contact surface between active solid and reactant gas and to the high resistance of the material to the diffusion of the reactant into the particle. Reducing the pellets size should improve the effectiveness factor but particle dimension cannot be too small,

since an excessive packing of the catalyst reduces the bed void fraction and increases pressure drops, which are usually important (2–3 barg).

- the high heat transport resistance of the packed bed, which causes a strong reduction of packed bed temperature and of methane conversion. As it is widely reported in literature, the random distribution of the pellets in the packed bed is characterized by a high void fraction in the zone near the tube wall, moreover which is a rigid boundary inhibiting lateral gas mixing. Consequently, in the zone near the hot tube wall the molecular conduction is the only heat transfer mechanism, causing a strong temperature drop immediately in the first layer of the gas-catalyst phase. Then, the heat flux encounters the high heat transport resistances of the gas mixture and of the catalytic pellets. Catalyst particles placed in the central zone of industrial large reformers usually do not work well since the temperature is too low for promoting the endothermic reactions.
- no water–gas shift conversion in the steam reformer. The very high temperature reached in the traditional steam reforming reactor changes the direction of the exothermic water–gas shift. Therefore, in the traditional steam reforming industrial plants, the reformer is followed by a water–gas shift reactor to reduce the carbon monoxide contents in the outlet stream.

5.2 Membrane Integration

Natural gas steam reforming process is an ideal candidate for hydrogen selective membrane integration tanks to its high reaction endothermicity and the fast kinetics leading to equilibrium condition inside traditional reactors.

The selective membrane integration allows a reaction product to be removed and equilibrium conditions are never reached. As a result, a specific feedstock conversion can be obtained at a lower operating temperature with the main benefits of a better heat integration, as the use of gas exhausts from a gas turbine as suggested by [3] or solar heated molten salts [4, 5].

By membrane integration, the design criteria of steam reforming plant have to be completely re-thought, carefully facing the membrane integration drawbacks, mainly coupling catalyst and membrane operating conditions and meeting a compromise optimization to promote both kinetics and permeability, without damaging the membrane which always requests a stringent thermal threshold.

Two plant configurations can be used:

1. the selective membrane is assembled directly inside the reaction environment, so that the hydrogen produced by reactions is immediately removed (MR concept);
2. the hydrogen selective membrane is assembled in separation units located downstream to reaction units. This configuration, called Reformer and Membrane Module (RMM), is composed by a series of reaction-separation modules.

Both configurations are described in the following paragraphs. MR is then modeled and simulated to assess the effects of operating conditions on reactor behavior, while for a description of a real RMM installation refer to Chap. 10.

5.2.1 Membrane Reactor Concept

A membrane reactor (MR) is a compact device in which a selective membrane is directly assembled. The simplest configuration is composed by two concentric tubes, where catalyst pellets are packed in the annular zone while the inner tube is the membrane itself, as shown in Fig. 5.8. Through the inner tube a sweeping gas (water steam) is sent, co-currently or counter-currently, to drag the hydrogen permeated. Of course, the membrane integration can also be made by assembling many smaller tubes, thus increasing the specific membrane surface per unit volume of reactor and consequently the overall hydrogen flow permeating. Two different zones can be recognized: the reaction zone, which is the annular section where catalyst is packed, and the permeation zone, where the sweeping gas is sent.

Many experimental works reported in the literature [7–15] attest the good performance in terms of natural gas conversion at much lower operating temperature than in traditional process (methane conversions up to 90–95% at 450–550°C vs. 850–1000°C).

Moreover, the MR configuration avoids packing catalyst in the central zone of the tubular reactor, thus improving the average catalyst effectiveness.

Furthermore, the lower operating thermal level required in MR leads to new procedures of process heat duty supply, avoiding the furnace and allowing hot fluids from other processes or solar thermal fluids [4, 5] to be used.

Membranes that can be used in MRs are:

1. dense metallic membranes, mainly Pd-based membrane, characterized by a very high hydrogen selectivity (almost infinite);
2. ceramic membranes, as silica membranes, with a stronger resistance to thermal stresses but lower hydrogen selectivity.

In modeling and simulating MR, Pd-based membranes are selected in this chapter to exploit their very high selectivity and because their development is at a pre-commercial status. For typical Pd-based selective membrane performance,

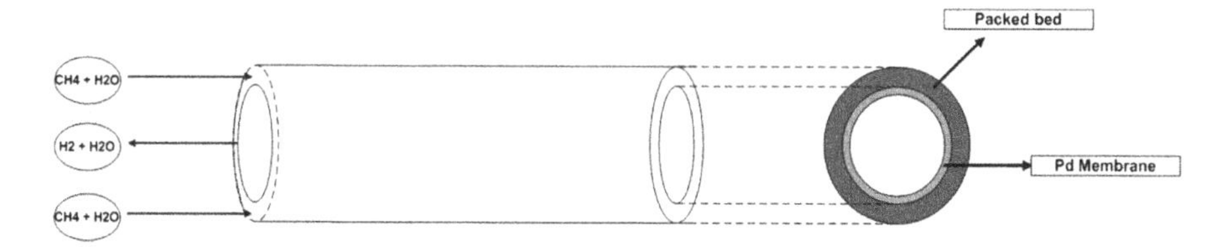

Fig. 5.8 Membrane reactor draft [6]

refer Table 2.3, where permeation data of different palladium–alloy membranes are reported.

The main criticism in membrane reactors is the need to impose the same operating conditions for membrane and catalyst, while:

- the catalyst has to operate at high temperature, to promote kinetics and avoid carbon coke formation reactions;
- the membrane must work below a thermal threshold (about 500°C for a reliable durability), to avoid stability problems due to stresses between active Pd layer and ceramic or SS support.

In Sect. 5.3 MR is modeled, simulated, and optimal operating conditions are assessed.

5.2.2 *Reformer and Membrane Modules*

In order to avoid the conflict of catalyst and membrane operating parameters, the selective membrane can be placed outside the reactor, in proper units located downstream. This is the concept of Reformer and Membrane Modules (RMM) steam reforming plant [3, 6, 16], shown in Fig. 5.9.

The feed is sent to a convective steam reformer where it is partially converted into hydrogen; then hydrogen is recovered through a Pd alloy membrane separation module, while the retentate is sent to the next step or recycled to the first module. By means of a heat recovery system, the operating temperature can be reduced to about 450°C before the membrane unit and again increased before the second reactor. It is possible to replicate the RMM until the desired natural gas conversion is achieved.

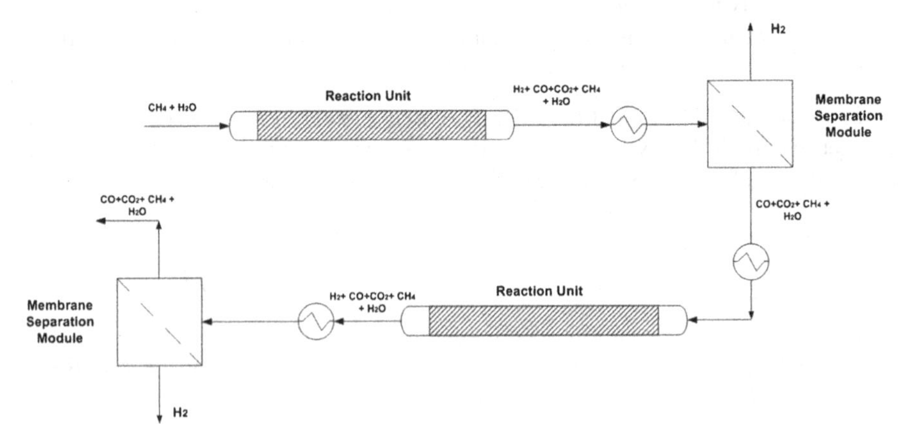

Fig. 5.9 Two reformer and membrane modules process scheme [6]

The main benefits of such a configuration are:

- the possibility to decouple reaction and separation operating conditions. Reforming and separation module temperatures can be optimized independently, both increasing methane conversion for each reaction step and membrane stability and durability.
- a simpler mechanical design of membrane tubes in comparison with the case of tubes embedded in catalyst; a simple "shell and tube" geometry can be selected for the tubular separation module.
- simplification of membrane modules maintenance and of catalyst replacement.

On the other hand, the main drawbacks in respect to MR configuration can be summarized as follows:

- the lower compactness of the plant: RMM configuration is composed by reactors, separation modules and heat exchangers between them, while in MR reaction and separation are performed in a single compact device.
- the membrane surface area required for the same methane conversion is greater in RMM configuration. This leads to a greater cost.

A RMM test plant with a capacity of 20 Nm^3/h of hydrogen and constructed in Chieti (Italy) by Italian process contractor Tecnimont-KT is presented in Chap. 10. Performance of such an innovative architecture is evaluated in a real pre-industrial application.

5.3 Membrane Reactor Performance

In order to simulate the behavior of an industrial membrane reactor, the model described in Sect. 4.2.2, i.e., a two-dimensional pseudo-homogeneous model, is fitted to natural gas steam reforming. Chemical–physical properties of reactions and components involved are taken from literature [17, 18] as functions of temperature, pressure, and mixture composition. The kinetic equations for the reactions scheme composed by 5.1, 5.2 and the overall reaction:

$$CH_4 + 2H_2O \Leftrightarrow CO_2 + 4H_2 \tag{5.5}$$

are taken from Xu-Froment [19].

The model is implemented in MatLab environment. The validation is made on the basis of industrial data sets [20] of a traditional steam reforming plant. Figure 5.10 reports the comparison between industrial and simulated data, attesting a good adherence [2].

Then, the dependence of membrane permeability on the temperature, given by Shu [7] and valid for a 20 μm Pd–Ag membrane, has been assumed and hydrogen down and upstream partial pressures (see Eq. 4.27) are fixed. The configuration depicted in Fig. 5.7 is assumed.

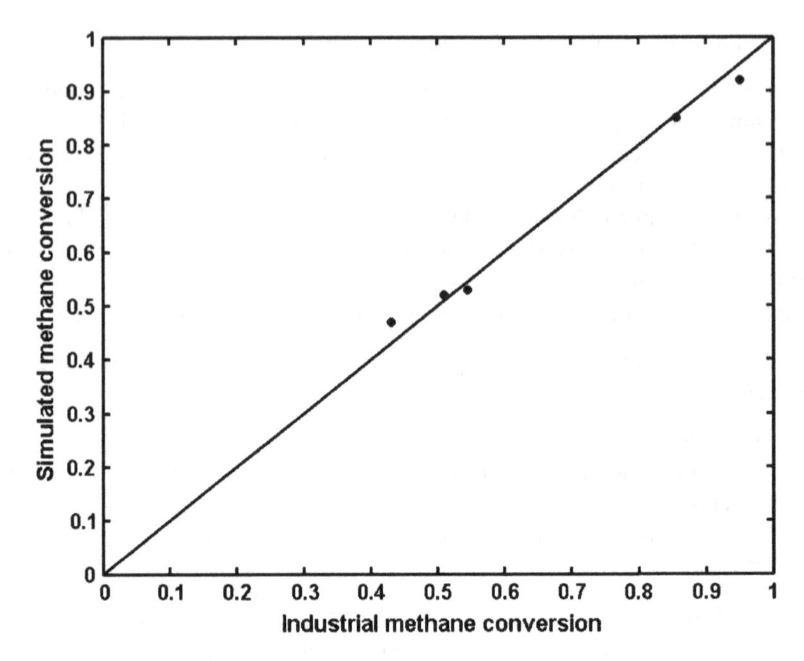

Fig. 5.10 Comparison between industrial methane conversions and simulation results

Table 5.1 MR operating conditions

MR length	Reactor diameter	Membrane diameter	Wall temperature
6 m	0.125 m	0.06 m	723–873 K
Permeation zone temperature	Sweep steam flow rate	Inlet methane flow rate	Steam to carbon ratio
773 K	4,000 h^{-1} (3 kmol/h)	800–4750 h^{-1} (2–12 kmol/h)	3–5
Membrane thickness			Pressure driving force
20 μm			5–20 bars

The operating conditions used in simulations are reported in Table 5.1 and are close to those ones of a potential industrial application; therefore reactor dimensions are similar to those of traditional reactors [20].

The following simulations are performed fixing the MR geometry and the permeation zone operating conditions and varying:

1. external wall temperature T_W;
2. inlet methane flow rate;
3. steam to carbon ratio;
4. pressure driving force between reaction and permeation zones, i.e. the operating pressure of reaction zone.

In evaluating the effect of the aforementioned operating conditions, a comparison with a traditional reactor with the same geometry is always made, to quantify the performance improvement by MRs.

Reactor performance is evaluated as:

- methane conversion
- pure hydrogen flow rate recovered through selective membrane (Nm^3/h)

Moreover, it is assumed that the outlet flow rate, composed by un-reacted methane, un-permeated hydrogen, CO, and CO_2, and therefore characterized by a great energy content, is sent to a gas turbine for electricity generation.

The simulations are performed for a single MR reactor.

5.3.1 Effect of Wall Temperature

As shown in Figs. 5.11 and 5.12, wall temperature of the external tube has a strong positive effect on MR performance for two main reasons:

1. reactions are promoted at higher temperatures;
2. membrane permeability increases at higher temperatures.

The benefits of integrating a selective membrane inside the reaction environment are self-evident: in the range of explored temperatures, the improvement of

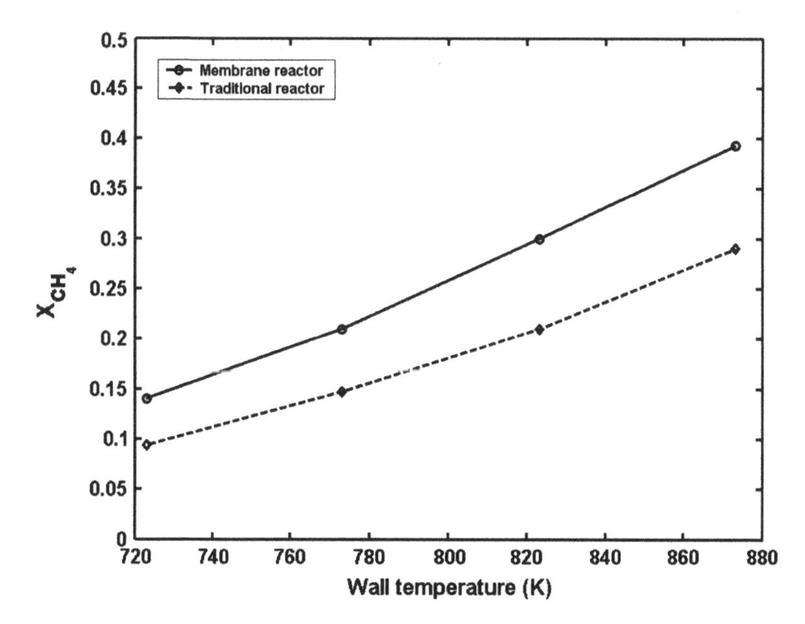

Fig. 5.11 Methane conversion in MR and TR versus wall temperature (reactor operating pressure = 15 bar, S/C = 3, inlet methane flowrate = 4 kmol/h)

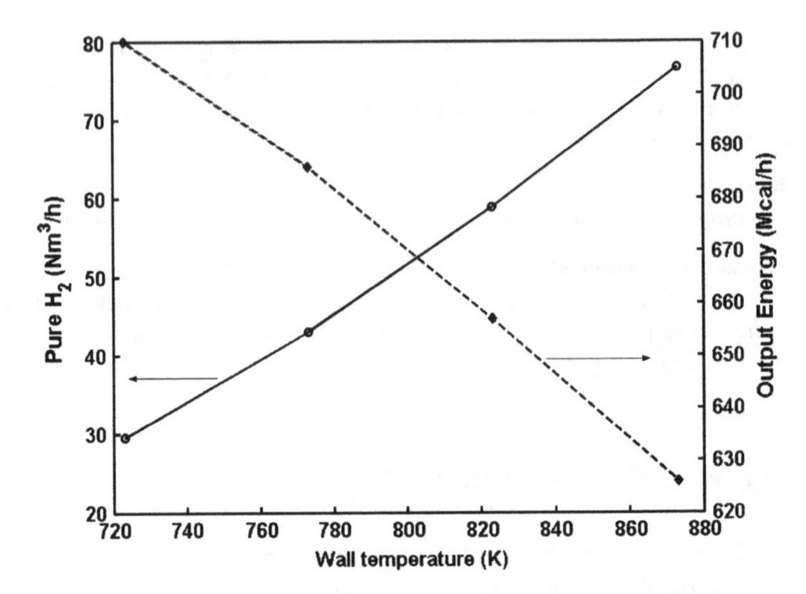

Fig. 5.12 Pure H_2 produced and recovered and energy of outlet stream (to be used for electricity generation) versus wall temperature (reactor operating pressure = 15 bar, S/C = 3, inlet methane flowrate = 4 kmol/h)

methane conversion in MR is within the range 38–46% in comparison with TR (Fig. 5.11).

Figure 5.12 shows that H_2 produced and recovered increases as well with the wall temperature but the output energy produced decreases since the amount of fuel sent to the gas turbine is lower.

On the other hand, increasing wall temperature leads to an increase of maximum membrane temperature as well. At $T_w = 873$ K, the membrane reaches a temperature of 830 K, while at 823 K the maximum temperature is 790 K. For this reason, the wall temperature, and consequently the heat flux supplied, has to be limited to avoid the overtaking of Pd-based membrane temperature threshold, fixed at 800 K with the present technology. Therefore, the wall temperature has to be maintained at values lower than 823 K.

5.3.2 Effect of Residence Time

Figure 5.13 shows the effect of residence time on MR behavior.

The residence time is calculated as the ratio between the catalyst packed volume and the inlet volume flow rate of reactants.

The advantage of MR in comparison with TR is greater at lower Gas Hourly Space Velocity (GHSV). The reason is that at high GHSV the residence time is too low compared to the characteristic time of permeation kinetics, and the hydrogen recovery percentage, i.e., the ratio between the pure hydrogen recovered

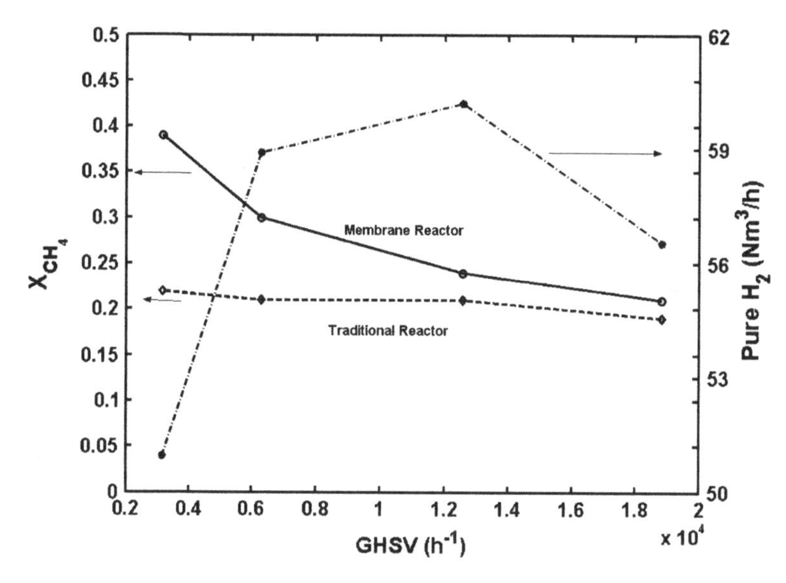

Fig. 5.13 Methane conversion in MR and TR and pure H_2 recovered through membrane versus inlet methane hourly space velocity (reactor operating pressure = 15 bar, S/C = 3, wall temperature = 823 K)

downstream and the total hydrogen produced, is low: 74.5% at 3,150 h^{-1} and 25.4% at 18,850 h^{-1}.

Pure hydrogen produced has not a monotonic trend, since GHSV has a double effect:

1. an increase of GHSV means an increase of reactant flow rate and thus an increase of hydrogen produced at a given conversion.
2. as shown in the figure, the increase of GHSV leads to a reduction of conversion and recovery ratio, which causes a reduction of hydrogen produced and permeated.

At low GHSV the first effect is leading, while at higher values of GHSV the second effect prevails.

Globally, if the GHSV value corresponding to the maximum pure hydrogen recovery is taken as optimal, the benefits of MR compared to TR are not enough to justify membrane integration (methane conversion improvement of 14.3%).

Therefore, the optimal value can be taken equal to 6,000–6,500 h^{-1}.

5.3.3 Effect of Steam-to-Carbon Ratio and of Pressure Driving Force

In Figs. 5.14 and 5.15, the effects of steam-to-carbon ratio and of reaction operating pressure on methane conversion and pure H_2 production are reported.

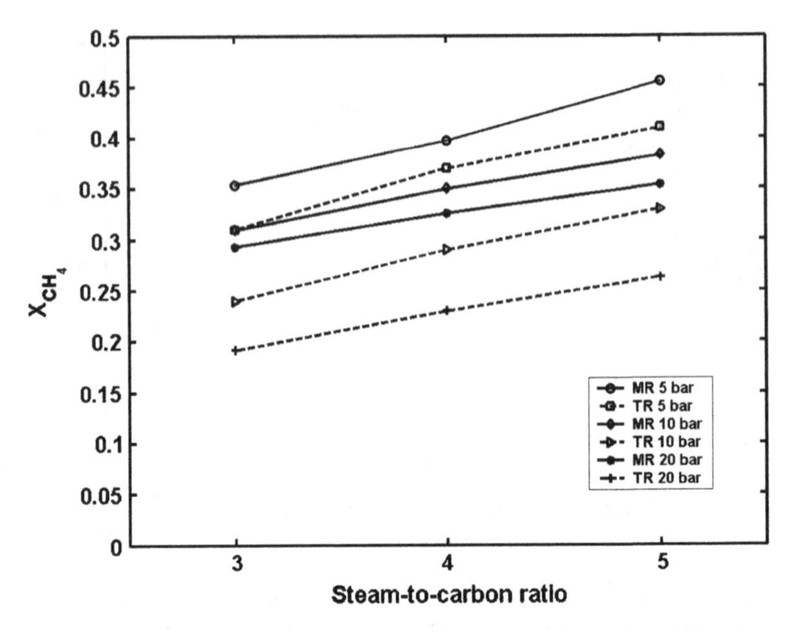

Fig. 5.14 Methane conversion in MR and TR versus steam-to-carbon ratio at different operating pressures (GHSV = 6,320 h^{-1}, wall temperature = 823 K)

High values of steam-to-carbon ratio improve reaction conditions, as the increase of methane conversion both for MR and TR shows (Fig. 5.14).

On the other hand, steam-to-carbon ratio has a double effect on permeation behavior:

1. high methane conversion, and consequently high hydrogen partial pressure in the reaction zone, should lead to an improvement of permeation rate;
2. a high flow rate of steam leads to a reduction of hydrogen partial pressure, making worse the permeation performance.

Globally, at low pressure the negative effect is prevailing, while at higher pressure the two effects are balanced (Fig. 5.15).

Also the pressure has a double effect on MR behavior:

1. it has a negative effect on reactions thermodynamics, leading to a reduction of equilibrium conversion;
2. it increases the pressure driving force, improving the membrane performance and consequently the pure H_2 recovery.

Altogether, it is a worth assessment that high pressure enhances the advantage of MR over TR, since the selective membrane works better.

A pressure of 10 bars and a steam-to-carbon ratio of 4 seem to be a good compromise for reaction and permeation performance.

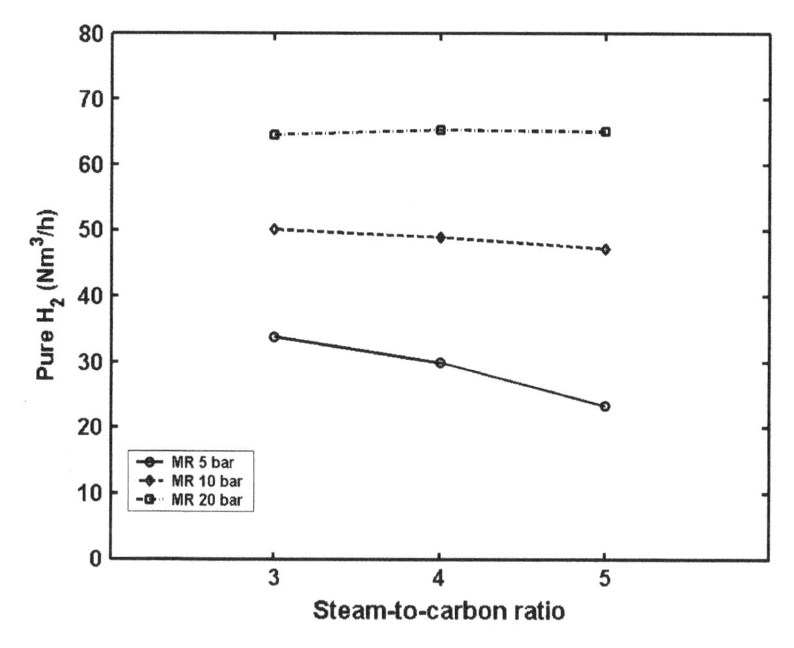

Fig. 5.15 Pure H_2 in MR versus steam-to-carbon ratio at different operating pressures (GHSV = 6,320 h^{-1}, wall temperature = 823 K)

5.3.4 MR Optimal Configuration

On the basis of previous simulations, the operating conditions reported in Table 5.2 are considered to be optimal.

Performing simulations at these conditions, the following outputs are obtained:

- methane conversion = 34.9% (improvement over TR is 22%)
- pure hydrogen produced = 49 Nm^3/h
- pure hydrogen recovered percentage = 40%
- output energy to turbo gas = 692 Mcal/h, able to produce 241 kW_{el} with the assumption of gas turbine electricity conversion efficiency of 0.3.

Figure 5.16 shows the temperature map inside MR reaction zone.

It has to be noticed the "cold spot" in the first reaction section, due to the high endothermicity of the steam reforming reaction, remarkable mainly where the reactant concentration is higher and reactions are faster.

Temperature radial gradient is never greater than 20 K, while in TR its values are much higher (up to 50–60°C).

Table 5.2 Operating conditions to be imposed for the steam reforming membrane reactor

Wall temperature	GHSV	Steam-to carbon ratio	Reaction zone pressure
823 K	6,320 h^{-1}	4	10 bars

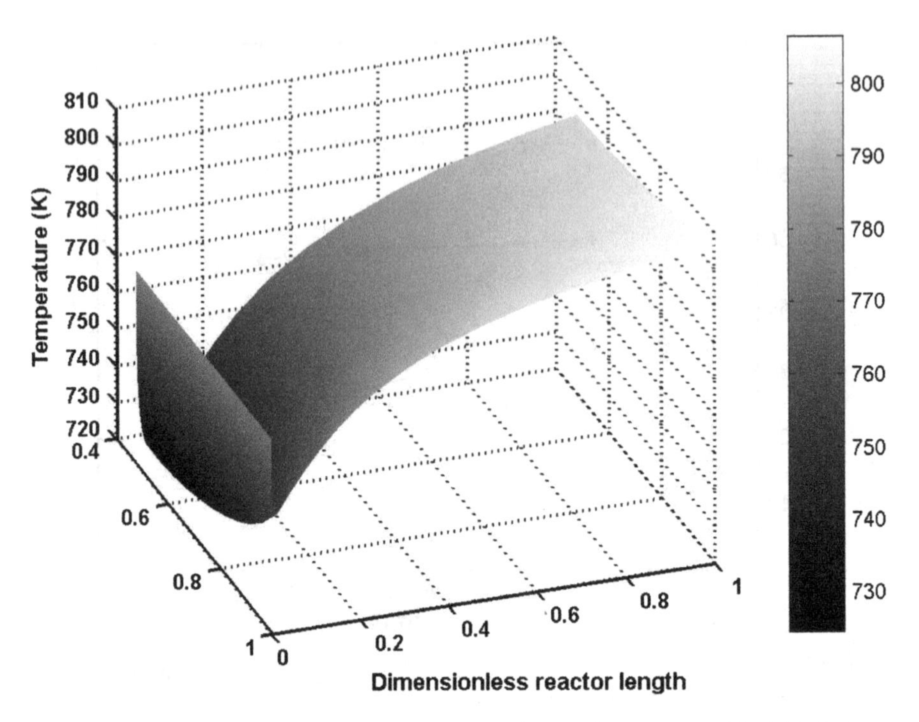

Fig. 5.16 Temperature in MR reaction zone at operating conditions reported in Table 5.2

On the basis of the single reactor productivity, 205 compact MRs are needed for an industrial plant able to produce 10,000 Nm^3/h of pure hydrogen without requiring any H_2 purification unit downstream the reactors system.

5.4 Future Perspectives

Membrane reactor technology is certainly a promising process intensification solution which should lead to crucial benefits toward a more efficient steam reforming process conduction.

The reduction of operating temperature drives the plant designers to find new and better solutions to supply heat duty to steam reforming reactors, exploiting, for example, heat streams from other parts of industrial plants.

Moreover, it has to be mentioned the possibility to couple the steam reforming process with gas-turbine flue gas [3] or with a solar source, considering that the lowered process thermal level is coherent with the most recent and innovative solar technologies, as molten salts Concentrating Solar Power plant [21–26], able to heat a molten salt stream up to 550°C by means of parabolic mirrors which concentrate solar energy on the focal line. The hot molten salt energy can be exploited to produce hydrogen by a steam reforming MR, as reported in [4].

On the other hand, the main criticism in steam reforming membrane reactors is the necessity to impose the same operating conditions for membrane and catalyst, whereas the catalyst should be at high temperature due to reactions endothermicity while membrane temperature must not exceed a threshold for assuring its stability. Considering that for the present technology the thermal limit for Pd-based selective dense membrane has to be imposed at values lower than 800 K, the MR performance is limited as well. Calculations reported in this chapter show that, at industrial operating conditions, methane conversion in MRs is limited at 35% about.

The technological development of Pd-based membrane is crucial for a wide diffusion of membrane reactors. R&D efforts have to be focused mainly on increasing membrane temperature threshold, on improving the adherence between the active layer and the support by new deposition and welding techniques, and on improvement of membrane reliability to produce objects suitable for industrial applications.

Future improvements of membrane performance would certainly promote the applications of MRs, but the current membrane state-of-the-art leads to the conclusion that RMM plant seems to be more suitable for industrial applications due to decoupling of reaction and separation operating conditions.

References

1. Vancini CA (1961) La Sintesi dell'Ammoniaca. In: Hoepli(ed.), Milan, Italy
2. De Falco M (2004) Pd-based membrane reactor: a new technology for the improvement of methane steam reforming process. Thesis, University of Rome "La Sapienza"
3. De Falco M, Barba D, Cosenza S, Iaquaniello G, Farace A, Giacobbe FG (2009) Reformer and membrane modules plant to optimize natural gas conversion to hydrogen. Asia Pacific J Chem Eng Memb React. doi:10.1002/apj.241
4. De Falco M, Basile A, Gallucci F (2010) Solar membrane natural gas steam reforming process: evaluation of reactor performance. Asia Pacific J Chem Eng Memb React 5:179–190
5. Giaconia A, De Falco M, Caputo G, Grena R, Tarquini P, Marrelli L (2008) Solar steam reforming of natural gas for hydrogen production using molten salt heat carriers. AIChE J 54:1932–1944
6. De Falco M, Iaquaniello G, Cucchiella B, Marrelli L (2009) Reformer and membrane modules plant to optimize natural gas conversion to hydrogen. In: Syngas: production methods, post treatment and economics, Nova Science Publishers Inc. New York. ISBN 978-1-60741-841-2
7. Shu J, Grandjean B, Kaliaguine S (1994) Methane steam reforming in asymmetric Pd and Pd–Ag porous SS membrane reactors. Appl Catal A General 119:305–325
8. Lin Y, Liu S, Chuang C, Chu Y (2003) Effect of incipient removal of hydrogen through palladium membrane on the conversion of methane steam reforming: experimental and modeling. Catal Today 82:127–139
9. Gallucci F, Paturzo L, Basile A (2004) A simulation study of steam reforming of methane in a dense tubular membrane reactor. Int J Hydrogen Energy 29:611–617
10. Oklany J, Hou K, Hughes R (1998) A simulative comparison of dense and microporous membrane reactors for the steam reforming of methane. Appl Catal A General 170:13–22
11. Madia G, Barbieri G, Drioli E (1999) Theoretical and experimental analysis of methane steam reforming in a membrane reactor. Canadian J Chem Eng 77:698–706

12. Yu W, Ohmori T, Yamamoto T, Endo E, Nakaiwa T, Hayakawa T, Itoh N (2005) Simulation of porous ceramic membrane reactor for hydrogen production. Int J Hydrogen Energy 30:1071–1079
13. Chai M, Machida M, Eguchi K, Arai H (1994) Promotion of hydrogen permeation on a metal-dispersed alumina membrane and its application to a membrane reactor for steam reforming. Appl Catal A General 110:239–250
14. Fernandez F, Soares A Jr (2006) Methane steam reforming modeling in a palladium membrane reactor. Fuel 85:569–573
15. Koukou M, Papayannakos N, Markatos NC (2001) On the importance of non-ideal flow effects in the operation of industrial-scale adiabatic membrane reactors. Chem Eng J 83:95–105
16. De Falco M, Barba D, Cosenza S, Iaquaniello G, Marrelli L (2008) Reformer and membrane modules plant powered by a nuclear reactor or by a solar heated molten salts: assessment of the design variables and production cost evaluation. Int J Hydrogen Energy 33:5326–5334
17. Perry R, Green D, Maloney J (1984) Perry's chemical engineers' handbook, 6th edn. McGraw Hill, New York
18. Reid R, Prausnitz J, Poling B (1988) The properties of gases and liquids, 4th edn. McGraw Hill, New York
19. Xu J, Froment G (1989) Methane steam reforming, methanation and water-gas shift: I. Intrinsic kinetics. AIChE J 35:88–96
20. Elnashaie S, Elshishini S (1993) Modelling, simulation and optimization of industrial fixed bed catalytic reactors, vol. 7 of Topics in Chemical Engineering, Gordon and Breach Science Publisher, New York
21. Winter CJ, Sizmann RL, Vant-Hull LL (1991) Solar power plants. Springer, New York
22. Mills D (2004) Advances in solar thermal electricity technology. Sol Energy 76:19–31
23. Herrmann U, Kearney DW (2002) Survey of thermal energy storage for parabolic trough plants. ASME J Sol Energy Eng 124:145–151
24. Pacheco JE, Showalter SK, Kolb WJ (2002) Development of a molten-salt thermocline thermal storage system for parabolic trough plants. ASME J Sol Energy Eng 124:153–159
25. Herrmann U, Kelly B, Price H (2004) Two-tank molten salt storage for parabolic trough solar power plants. Energy 29:883–893
26. Kearney D, Herrmann U, Nava P, Kelly B, Mahoney R, Pacheco J et al (2003) Assessment of a molten salt heat transfer fluid in a parabolic trough solar field. ASME J Sol Energy Eng 125:170–176

Chapter 6
Autothermal Reforming Case Study

Paolo Ciambelli, Vincenzo Palma, Emma Palo and Gaetano Iaquaniello

6.1 Process Description

There are three primary techniques used to produce hydrogen or syngas from hydrocarbon fuels: steam reforming (6.1), partial oxidation (6.2), autothermal reforming (ATR) (6.3) [1–3].

$$C_mH_n + mH_2O = mCO + (m + n/2)H_2 \tag{6.1}$$

$$C_mH_n + \frac{1}{2}mO_2 = mCO + \frac{1}{2}nH_2 \tag{6.2}$$

$$\begin{aligned} C_mH_n + xH_2O + yO_2 = aCO + (m - a)CO_2 + (x + 2y - 2m + a)H_2O \\ + (n/2 - 2y + 2m - a)H_2 \end{aligned} \tag{6.3}$$

Each of this process shows several advantages and challenges. Steam reforming is an endothermic reaction, therefore, it requires an external heat source but it does not require oxygen. Furthermore, it produces the reformate with the highest H_2/CO ratio (3) and is the typical preferred process employed in industry for hydrogen production. In partial oxidation, using a substoichiometric amount of oxygen or air, a partial or incomplete combustion of the fuel occurs. The reaction can be carried out in the presence or the absence of a catalyst. In the non-catalytic process higher temperatures must be reached to obtain complete CH_4 conversion, however, the advantage is that fuel components such as sulphur compounds do not need to be removed, thus much heavier petroleum fractions can be processed making this

P. Ciambelli (✉) and V. Palma
Department of Chemical and Food Engineering, University of Salerno, 84084 Fisciano SA, Italy
e-mail: pciambelli@unisa.it

G. Iaquaniello and E. Palo
Tecnimont KT, S.p.A, Viale Castello della Magliana 75, 00148 Rome, Italy
e-mail: Iaquaniello.G@tecnimontkt.it

M. De Falco et al. (eds.), *Membrane Reactors for Hydrogen Production Processes*,
DOI: 10.1007/978-0-85729-151-6_6, © Springer-Verlag London Limited 2011

reaction more attractive for processing diesel, logistic fuels and residual fractions cuts. The reaction is characterized by a H_2/CO ratio of 2 which may be suitable for Fischer–Tropsch synthesis.

Autothermal reforming is a stand alone process which combines partial oxidation and steam reforming in a single reactor. The ATR process was first developed in the late 1950s by Topsøe mainly for synthesis gas production in ammonia and methanol plants [4]. The hydrocarbon feedstock is reacted with a mixture of oxygen or air and steam in a sub-stoichiometric flame. In the fixed catalyst bed the synthesis gas is further equilibrated. The composition of the product gas is determined by the thermodynamic equilibrium at the exit pressure and temperature, through the adiabatic heat balance based on the composition and flows of the feed, steam and oxygen added to the reactor. The produced synthesis gas is completely soot-free [5].

An ATR reactor consists of a thermal or combustion zone and a catalytic one. In the first zone part of the feed is partially oxidized, generating the heat needed to drive the downstream steam reforming reactions in the catalytic zone. By proper adjustment of oxygen to carbon and steam to carbon ratios the partial combustion in the thermal zone supplies the heat for the subsequent endothermic steam reforming reaction [6]. This characteristic makes peculiar the temperature profile in an ATR reactor. Indeed, the catalyst bed-temperature rises rapidly near the reactant inlet end of the bed, and then gradually decreases, as result of a sequential reaction mechanism involving the initial exothermic oxidation reactions of methane, followed by the strongly endothermic steam reforming reaction [7–15], as well as the mildly exothermic water gas shift reaction (6.4):

$$CO + H_2O = CO_2 + H_2 \tag{6.4}$$

Unlike the steam reformer the autothermal one requires no external heat source and no indirect heat exchangers. This makes autothermal reformers simpler and more compact than steam reformers, resulting in lower capital cost. Furthermore, autothermal reformers typically offer higher system efficiency than partial oxidation systems, where the excess heat is not easily recovered. The autothermal reforming reaction is carried out in the presence of a catalyst, which controls the reaction pathways and thereby determines the relative extents of the oxidation and steam reforming reactions. Therefore, in order to achieve the desired conversion and product selectivity an appropriate catalyst is essential.

Catalyst formulations for ATR are strongly dependent on the selected fuel and the operating temperature. A good catalyst has to be simultaneously active for hydrocarbon oxidation and steam reforming reactions, to be robust at high temperature for extended times of operations and to be resistant to sulphur poisoning and carbon deposition, especially in the catalytic zone that runs oxygen limited [16] and in particular, with hydrocarbon feed heavier than methane. Finally, it must be resistant to intermittent operation and cycles, especially on start-up and shut-down steps. The preferred catalyst for ATR is a low loaded nickel-based catalyst supported on alumina (α-Al_2O_3) and magnesium aluminate spinel ($MgAl_2O_4$) since spinel has a higher melting point and in general a higher thermal strength and stability than alumina [2]. Other supports such as CeO_2, ZrO_2 or CeO_2–ZrO_2, γ-Al_2O_3 have been also suggested [17–23]. Pt-, Rh-, Ru- and Pd-based catalysts are also used for ATR, in particular with feed different from methane [24–39]. Alternative catalyst formulations for methane ATR based on bimetallics have been studied [40, 41] because the activity of nickel catalysts can be increased by the addition of low contents of noble metals [42–47]. Furthermore, in recent years, research into catalysts for ATR of hydrocarbons has paid considerable attention to systems with a perovskite structure of general formula ABO_3 [48–52], which are strong candidates as precursors of reforming catalysts due to the possibility of obtaining well-dispersed and stable active metal under the reaction conditions.

For small scale units providing hydrogen for fuel cells the choice of the optimal technology may be dictated by parameters such as simplicity and fast transients response, in particular for automotive applications. For these reasons, even if the steam reforming process may be most energy efficient in industrial process, catalytic partial oxidation or autothermal reforming using air may be the preferred choice for small scale or distributed hydrogen production [53]. As discussed by Ahmed and Krumpelt [54], the lower temperature provides many benefits such as less complicated reactor design and lower reactor weights, due to less thermal integration, less fuel consumption during the start-up phase, wider choice of materials thus reducing the manufacturing costs in small scale applications. The optimization of the structure of the catalyst support is also important. Due to the severe operating conditions (high temperature and high flow rates), typical of short contact time reactions, heat and mass transfer properties are expected to play a decisive role in the behaviour of the reactor [55]. Structured monolith or foam-shaped catalysts show several advantages over pellet catalysts, the most important one being the low-pressure drop associated with the high flow rates that are usual in environmental applications. Since structured catalysts can operate at significantly higher space velocities than pellets bed reactors, the consequent reduction in reactor size saves both cost and weight. A secondary benefit of having a smaller reactor to be heated is that a rapid thermal response to transient behaviour can be achieved. Whether the application is a load-following stationary fuel cell or an on-site hydrogen generator stepping up from standby mode to full operation, the reactors need to be able to respond quickly to changes in temperatures and flow rates [56]. Furthermore, in traditional fuel processor, after the reforming reactor there is a section devoted to syngas cleaning from CO which, in particular when the end-use technology is a PEM fuel cell, is a poison. Thus, to obtain a pure hydrogen stream several further stages such as water gas shift, methanation, CO preferential oxidation, membrane separation are necessary. However, when the final target is a reduced fuel processor size, a possible solution is the integration of the syngas production and purification units in the same reactor, employing a H_2 membrane selective. Applying an H_2 membrane to an ATR reactor can simplify the process for pure hydrogen production, and in the meantime can promote methane conversion into hydrogen. As CH_4 is a stable hydrocarbon, high reaction temperatures, around 800°C, are required for the endothermic reforming reaction. If H_2 is selectively and continuously removed from the reaction system, the equilibrium limitations of a conventional reactor can be circumvented, shifting the thermodynamic product distribution of steam reforming and CO shift reactions to the product sides, and highly efficient conversion of CH_4 to H_2 and CO_2 can be attained at a much lower operating temperature than in a conventional reactor [57]. However, the typical ATR temperature profile with large temperature excursions close to the catalyst bed inlet makes its integration with a Pd-based membrane very difficult, due to poor thermal–mechanical stability of these membranes.

This chapter reports on the integration of a kW-scale CH_4-ATR reactor, developed at the University of Salerno, with a membrane for hydrogen separation. The chapter focuses attention on the careful and proper procedures and operations to be implemented in order to carry out the integration avoiding the membrane damage. Over the last years several activities relevant to methane autothermal reforming were carried out at the University of Salerno, in the framework of the Italian FISR Project" Idrogeno puro da gas naturale mediante reforming a conversione totale ottenuta integrando reazione chimica e sepatrazione a membrana" in collaboration with other universities and Tcnimont KT of Rome. The developed ATR reactor, up to 5 m^3 (stp)/h of hydrogen capacity, has a very fast start-up phase (lower than 3 min) and stable operation in a wide range of conditions, and is self sustained due to thermal integration achieved with two heat exchangers for feed preheating by sensible heat of the reaction products. By this integration, it is operated without any external heat source, since the liquid water fed to the reactor is vaporized inside of it [58–60]. The reactor, loaded with structured catalyst in form of either noble metals- or nickel-based honeycomb monolith or open cells foam proved how much mass and heat transfer may determine a good performance or high hydrocarbon conversion [61–63]. Further improvements were obtained by changing the flow direction through the catalytic bed, evidencing how a radial flow may assure a more flat temperature profile along the catalytic bed rather than a traditional axial flow, thus being more proper for the integration with a membrane [64, 65]. A further flattening of temperature profile was obtained employing high thermal conductivity (such as SiC) supports. The reactor was also successfully integrated with a water gas shift stage for preliminary CO abatement in the exhaust stream. Also in this case using a structured catalyst enables for high CO conversions with a thermal efficiency of about 72% [66]. Moreover, we found that higher CO conversion is achieved with radial more than axial flow and that in the former case a centrifugal configuration has to be preferred to centripetal configuration [67].

6.2 Membrane Integration

6.2.1 *State of the Art*

Few studies are reported in literature dealing with the integration of an autothermal reforming reactor with a membrane for hydrogen separation, and most of them are simulation studies.

Lattner and Harold [68] carried out a comprehensive modelling study comparing conventional and membrane reactor fuel processors in terms of efficiency (LHV basis) and reactor volume, considering three process configurations: (i) 'conventional', consisting of a *n*-tetradecane ATR reactor, followed by a cooled WGS reactor and a preferential oxidation (PROX) reactor; (ii) 'palladium membrane reactor', consisting of a *n*-tetradecane ATR reactor integrated with a dense Pd membrane; (iii) 'dense oxide membrane reactor', consisting of a *n*-tetradecane ATR reactor integrated with a dense ceramic membrane. In each scheme the ATR reactor is integrated into an overall process model including fuel cell (50 kW), heat integration exchangers, and water recycle. It was found that the Pd membrane reactor has several advantages with respect to the 'conventional' configuration such as: (i) a 20% reduction in the fuel processor volume; (ii) a small improvement in the overall system efficiency (40.7% against 39.7%) due to elimination of the hydrogen losses in the PROX step; (iii) the use of steam as sweep gas, providing a pre-humidified anode feed gas, which can be a requirement in many PEM fuel cell designs. However, in order to maximize the fuel processor efficiency, high temperature (900°C) should be employed, resulting in membrane detriment. This issue, as well as that relevant to the high Pd cost may be overcome by using oxide-based membranes which showed reasonable performance in terms of productivity. A following modelling study by Lattner and Harold [69] focused attention on the comparison between three methanol fuel processor integrated with a PEM fuel cell (50 kW): (i) steam reformer followed by a PROX reactor; (ii) autothermal reformer followed by a PROX reactor; (iii) autothermal reformer integrated with a Pd membrane with countercurrent steam sweep gas. For each configuration the efficiency and the fuel processor volume were evaluated employing the reaction kinetics of a $CuO/ZnO/Al_2O_3$ catalyst. Both steam reformer and ATR reactor followed by PROX stages showed the same efficiency (50%, LHV basis), because the advantage of the ATR system relevant to the absence of external heat sources is counterbalanced by the higher steam reformer hydrogen concentration in the reformate. The slightly lower efficiency of the ATR membrane reactor is due to the additional steam generation required for the sweep gas, while the main advantage is the reduction in fuel processor volume due to the elimination of PROX reactor (12.7 litres against 21.4 litres of ATR plus PROX). The feasibility of a packed bed membrane reactor for the ATR of methane was also investigated by Tiemersma et al. [70], by developing a two-dimensional, pseudo-homogeneous reactor model to calculate the axial and radial temperature profiles and product distribution in the reactor. Due to the large temperature peak at the reactor entrance, which can be detrimental for life and stability of Pd membrane, two different operation modes were investigated to reduce these effects: wall-cooled operation with a high sweep gas rate, and staged oxygen feed, distributed along the length. In the first case temperature gradients along the membrane can be avoided, however, a decreased hydrogen removal rate as well as higher CO concentration in the reactor exhaust were observed. On the other hand the staged oxygen feeding provided for a reduction in temperature excursions, but a significant lower average reactor temperature was observed with a reduction in methane conversion and system performance. The major disadvantage of this operation is the more complex reactor design and the loss in energy efficiency between the stages. The authors assessed the feasibility of the Pd membrane integration in the reactor, but the membrane must be positioned outside the inlet region with large temperature gradients.

A detailed simulation work dealing with the evaluation of the influence of several parameters such as the flowing mode and the rate of sweeping gas, the inlet rate of CH_4, and the inlet ratio CH_4 to H_2O was carried out by Feng et al. [71]. A Ni/Al_2O_3 catalyst was considered for the simulation and a H_2-membrane module, consisting of 100 membrane tubes, was applied from 0.2 m away from the inlet of the reactor in order to avoid too high temperature. The counter-current flowing mode of sweep gas (steam) was found more suitable with respect to the co-current, due to a larger average driving force for hydrogen permeation all along the reactor length, thus resulting in higher molar flow rate of separated H_2. At increasing inlet flow rate of CH_4 to the reactor the average temperature of both reaction and permeate side increases, as well as the molar flow rate of H_2 in the reaction side, the average driving force of H_2 permeation and the production rate of separated H_2. However, in these conditions the fraction of hydrogen recovered (defined as the ratio of separated H_2 to the total useful products of CO and H_2) and the CH_4 conversion decrease. Thus, an optimal inlet rate of CH_4 has to be found in order to maximize either the production rate of separated H_2 per overall CH_4 either the thermodynamic efficiency. At decreasing CH_4/H_2O inlet ratio more steam is added in the feed, and the H_2 molar flow rate in the gas mixture is higher than that at higher CH_4/H_2O ratio. However, more inlet steam leads to a lower partial pressure of H_2 in the reaction side, thus lowering the average driving force of hydrogen permeation. This also corresponds to a decrease in thermodynamic efficiency. Finally, by increasing the rate of sweeping gas, higher production rate of separated H_2, higher fraction of H_2 recovered, and higher methane conversion were obtained. However, these results do not mean that the increase of the rate of sweeping gas is always necessary, but that for a fixed inlet rate of CH_4 there is a rate of sweeping gas that is just sufficient to separate the produced hydrogen maximally, while a further increase will result in negligible improvements.

Other reported configurations of ATR membrane reactors are those involving both O_2 and H_2 permeable membranes [72, 73] (the presence of O_2 permeable membrane enables for the injection of oxygen from air at several points over the length of the reactor in order to reduce temperature hot-spots as also reported by Tiemersma [70]), or fluidized bed reactors [74–81] or a reactor proposed by Simakov and Sheintuch [82]. The latter is assumed to be composed of two concentric cylinders: a fixed bed for steam reforming reaction which contains one or several Pd membranes, surrounded by an outer tube packed with a catalyst for oxidation reaction. The main advantage of fluidized bed reactors is the thermal uniformity in the reaction chamber which is advantageous for maximising membrane utilization as well as for minimizing thermal stresses in the membranes. The fluidized bed also provides a greater flexibility to achieve a high membrane surface area per unit reactor volume compared to packed beds. However, some challenges are catalyst attrition/erosion, a more complex design/scale-up/construction process.

In the example reported by Elnashaie et al. [74–76], Pd-based hydrogen permselective membrane tubes and dense perovskite oxygen membrane tubes inside a heptane reformer were considered. The fluidized bed was circulating, thus the eventually spent catalyst was carried out of the reformer with the exit gas stream, regenerated along the exit line, separated and recycled. The fast flow of catalyst makes the effect of carbon deposition on the catalyst activity negligible. In the example reported by Patil et al. [77] the oxidation section and the reforming/shift section were separated, with the former containing a O_2 permeable membrane and the latter containing a H_2 permeable membrane. In the oxidation section CH_4 is partially oxidized in order to achieve the high temperatures required for O_2 permeation through the perovskite membranes and to simultaneously preheat part of the CH_4/steam feed. The preheated feed is mixed with additional CH_4 and steam and fed to the reforming/shift section, where CH_4 is completely reformed to CO_2 and H_2. The reactor modelling demonstrated that autothermal operation and effective temperature control in both reactor sections can be achieved along with high CH_4 conversion and H_2 yields. Moreover, an increase in the superficial gas velocities leads to an increase in the power output, however at the expense of higher CH_4 and H_2 losses in the reactor exhaust.

6.2.2 Experimental Apparatus

The laboratory plant employed for the catalytic activity tests in the presence of a membrane integrated in the autothermal reforming reactor is reported in Fig. 6.1.

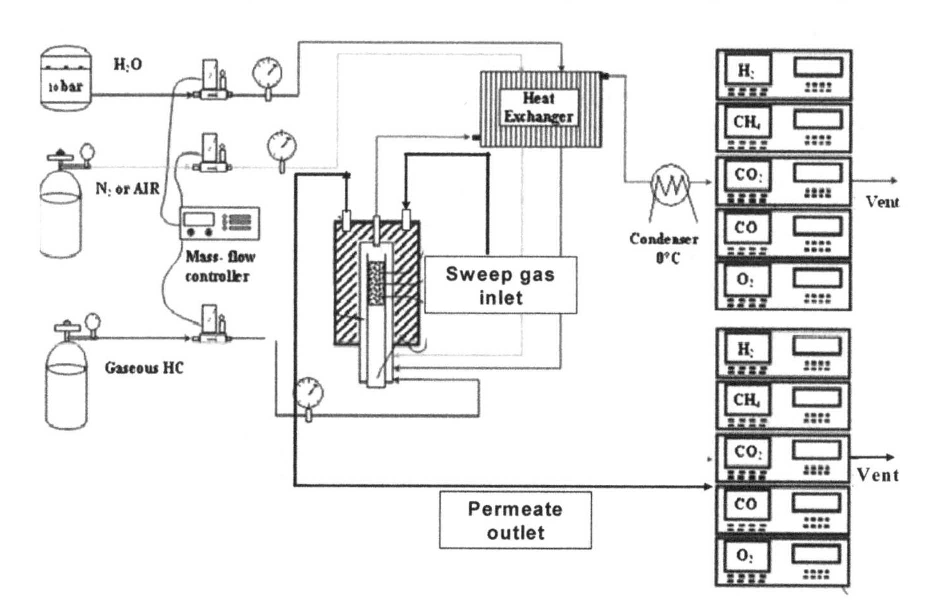

Fig. 6.1 Laboratory apparatus plant

The main part of the plant is represented by the ATR reactor (36 mm i.d.). It mostly consists of a lower section where, only during the start-up phase, does methane react with air at a fixed O_2/CH_4 ratio, and an upper section where reforming reactions occur in the presence of the catalyst. Due to the specific O_2/CH_4 ratio in the start-up phase, in the lower section, owing to the exothermicity of the hydrocarbon oxidation, heat is released and transferred to the reforming section heating the catalytic bed up to the ATR catalyst threshold temperature. Methane and air were fed at the bottom of the reactor and premixed in a mixing chamber; a SiC foam is placed at the exit of the mixing chamber in order to obtain a well distributed and homogeneous flame. In the start-up phase the ignition of methane-air mixture is induced by a voltaic arc between two spark plugs placed on the surface of the SiC foam. The catalyst bed (around 70 cm^3) is located in the reforming section supported by a metallic gauze. Water is fed to the reactor at the bottom of the metallic gauze. The temperature inside the reactor is monitored by several thermocouples, one of which is located on the SiC foam, while other thermocouples are placed in the catalytic section to provide the temperature profile in the reactor's axial direction. Furthermore, additional thermocouples are placed to monitor the temperature of preheated water and air, of the exhaust stream before and after thermal exchange, and of the stream in the premixing chamber. A differential pressure sensor monitors the pressure drop across the reactor, and the coke formation. The outside shell of the reactor is thermally insulated to reduce heat loss. The ATR reactor is integrated with two heat exchangers to preheat the air and the water fed to the reactor by the exhaust stream. At the reactor outlet a fixed fraction of exhaust stream is sent to the analysis section through a cold trap to remove water. Two sets of analyzers monitor simultaneously gaseous component concentrations on both permeate and retentate side. In particular, CH_4, CO and CO_2 concentrations were monitored with an on line NDIR analyzer, for the analysis of O_2 a continuous paramagnetic analyzer was employed, H_2 was analysed with a thermo conductivity analyzer.

The tests were carried out with a cylindrical membrane made up of Al_2O_3 (Inocermic GmbH), the length of the membrane module was around 10 cm (Fig. 6.2).

The module was activated with the following thermal pre-treatment:

1. Ramp temperature 1°C/min to 105°C;
2. Isotherm at 105°C for 1 h;

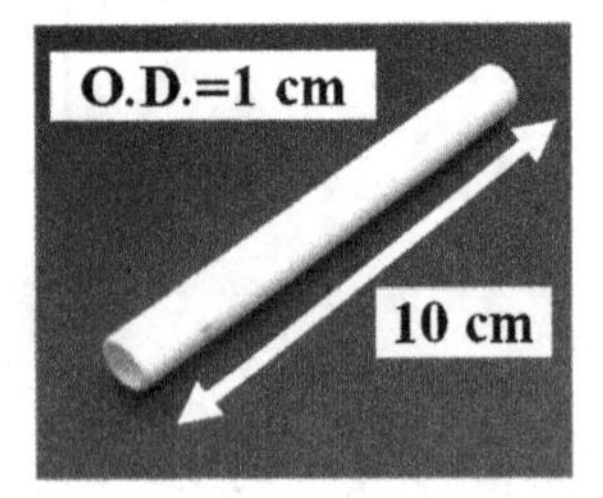

Fig. 6.2 Membrane module

3. Ramp temperature 1°C/min to 140°C;
4. Isotherm at 140°C for 1 h;
5. Ramp temperature 1°C/min to operating temperature.

This procedure was carried out by flowing an inert gas on both tube and shell sides of the membrane.

In order to realize the integration of the structured ATR reactor with the membrane high attention must be paid to the choice of the catalyst structure. For this reason a foam structured support was chosen in order to enable the gas flowing through the catalyst to the membrane surface, due to its porous network (Fig. 6.3a). Furthermore, the catalyst was annular shaped to realize the proper insertion of the membrane (Fig. 6.3b).

The choice of the catalyst structure was optimized. Indeed, our previous activity was relevant to the optimization of the temperature profile along the catalytic bed, in order to ensure the lowest strain for the operation of the membrane [64, 65]. We found that a high thermal conductivity material, such as silicon carbide, along with a radial flow geometry may lead to a more isothermal profile along the catalytic bed.

In this more innovative assembly, five elements of catalyzed foams were collected as reported in Fig. 6.4.

An alumina adhesive layer is placed on both ends of catalytic bed, in order to provide a radial flow for gases coming from the mixing chamber and the membrane tube is inserted inside the catalytic bed. Furthermore, in order to allow simultaneously the sweep gas inlet and the permeate outlet, a stainless steel component (Fig. 6.5) is inserted at the top of the reactor.

It contains two internal tubes; the sweep gas enters in one of them, flows through a coaxial tube located inside the membrane and allows the transport of the permeate gas through the second internal tube.

Since the membrane and the catalytic bed located around the membrane are inserted upside down into the reactor body (Fig. 6.6) it was necessary to use an

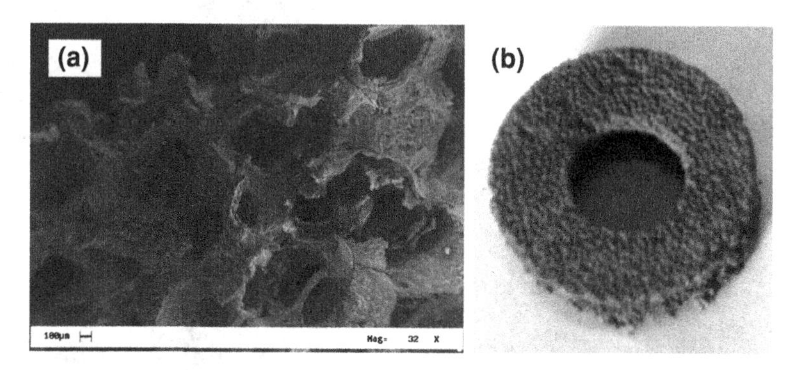

Fig. 6.3 SEM image of foam porous network (**a**) and catalyst shape for the insertion of the membrane (**b**)

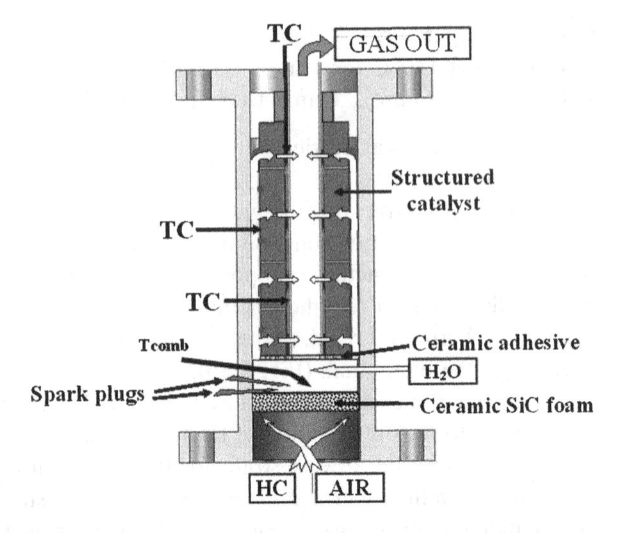

Fig. 6.4 Catalyzed foams radial assembly

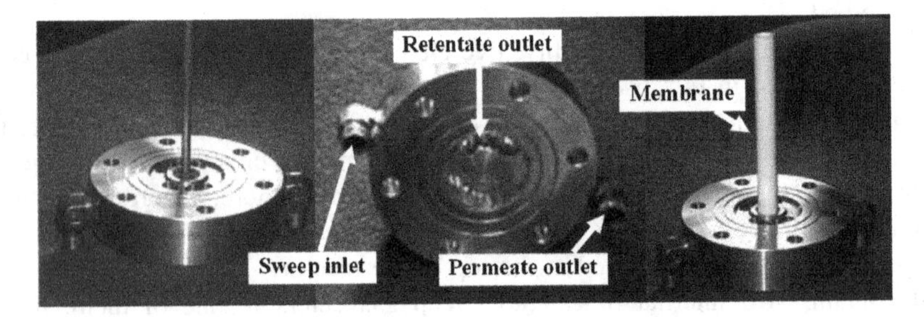

Fig. 6.5 Stainless steel component

Fig. 6.6 Insertion of the membrane and catalytic bed inside the reactor body

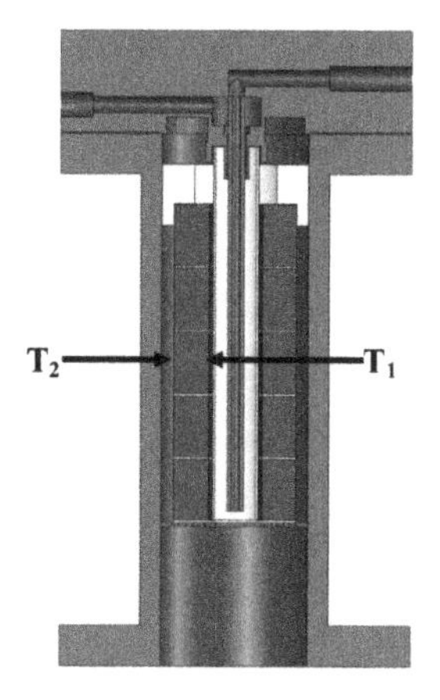

Fig. 6.7 Location of thermocouples inside the catalytic bed for test with membrane integration

adhesive for bonding the ceramic membrane and catalyst to the reactor stainless steel and avoiding any flow of the retentate gas in the permeate side.

Two thermocouples were inserted in the reactor to monitor the temperature at the inlet (T_2) and outlet (T_1) side of the catalytic bed, with the T_1 very close to the membrane surface (Fig. 6.7).

Furthermore, due to the absence of external heat sources, a metallic-shielded heating tape controlled by a PID temperature controller was wrapped all around the external surface of the reactor body, in order to carry out the thermal pre-treatment of the membrane.

6.3 Membrane Reactor Performance

The performance of the membrane reactor was evaluated in the following operating conditions:

1. molar feed ratio $O_2/CH_4 = 0.56$;
2. molar feed ratio $H_2O/CH_4 = 1.5$;
3. absolute pressure ranging between 1 and 5 atm;
4. gas hourly space velocity (GHSV) = 12,000 h^{-1}

where GHSV was defined as the ratio between the total gaseous flow rate (STP) fed to the reactor and the volume of the whole catalytic bed. In these operating conditions, derived from a previous optimization [64, 65], the maximum temperature reached along the catalytic bed was about 505°C, a suitable value for avoiding any damage to the membrane.

The reactor start-up followed the step of membrane thermal pre-treatment (Fig. 6.8): the electrical heating tapes were turned off and the reactor was fed with a mixture of methane, air and nitrogen in order to heat the combustion zone and the catalyst up to the operating conditions in two different steps. In the first step methane combustion occurred homogeneously on the silicon carbide foam,

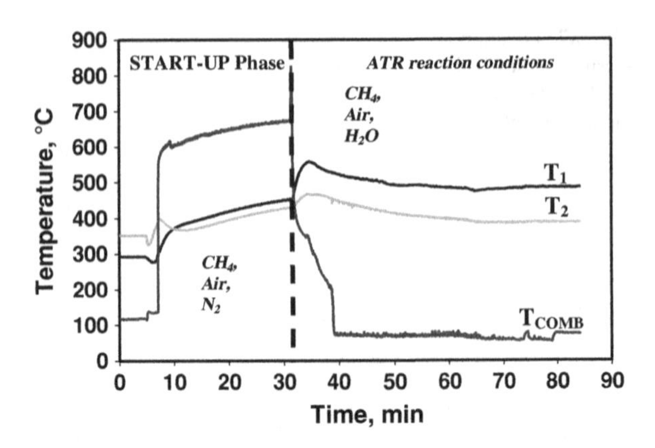

Fig. 6.8 ATR membrane reactor start-up phase

and heat transfer caused the increase in the temperature of the system catalyst-membrane, heated under nitrogen flow. When the temperature of the membrane reached around 400°C, the nitrogen flow was stopped, water was added to the reaction mixture and the operating conditions were moved to the desired ones. As a consequence, the temperature of the SiC foam suddenly decreased and a slight increase in the temperature of the catalyst-membrane occurred, without rising above 500°C.

The results of catalytic activity tests carried out with the INOCERMIC membrane at 1 atm are reported in Fig. 6.9 in terms of CH_4, CO, CO_2, H_2 concentration (vol%, dry basis) at both retentate and permeate sides as function of time. During the start-up phase only CO_2 was detected at the reactor outlet at the retentate side, due to the methane total combustion, while at the permeate side any gaseous component was detected. After the reactor start-up the more reducing conditions allowed to obtain a mixture of CH_4, CO, CO_2 and H_2 at both retentate and permeate sides, with a remarkable amount of H_2 with respect to the other components. The H_2 concentration at the permeate side was about 2 vol%, while the temperature close to the membrane surface was about 490°C.

The progressive increase of the absolute pressure led to a progressive decrease in the H_2 concentration and CH_4 conversion on the retentate side (Fig. 6.10), due to the harmful effect of pressure on the reforming reactions. In spite of this, on the permeate side, an increase of the permeated hydrogen as well as the other species concentration occurred, with the product distribution remaining almost the same.

The obtained data were elaborated with the Knudsen (or free-molecule) diffusion model, according to which the flux of a gaseous species through a membrane is directly proportional to its partial pressure at retentate side and has an inverse dependence on the square root of its molecular weight [83]. Thus, in order to validate the model of Knudsen diffusion through the membrane, the ratios H_2/CH_4, H_2/CO and H_2/CO_2 were evaluated on the permeate side and compared with the expected ones (Fig. 6.11).

It can be observed that for CH_4 the experimental values are close to the expected ones for Knudsen diffusion at each operating pressure. This should be ascribed to a Knudsen diffusion. However, besides Knudsen diffusion another

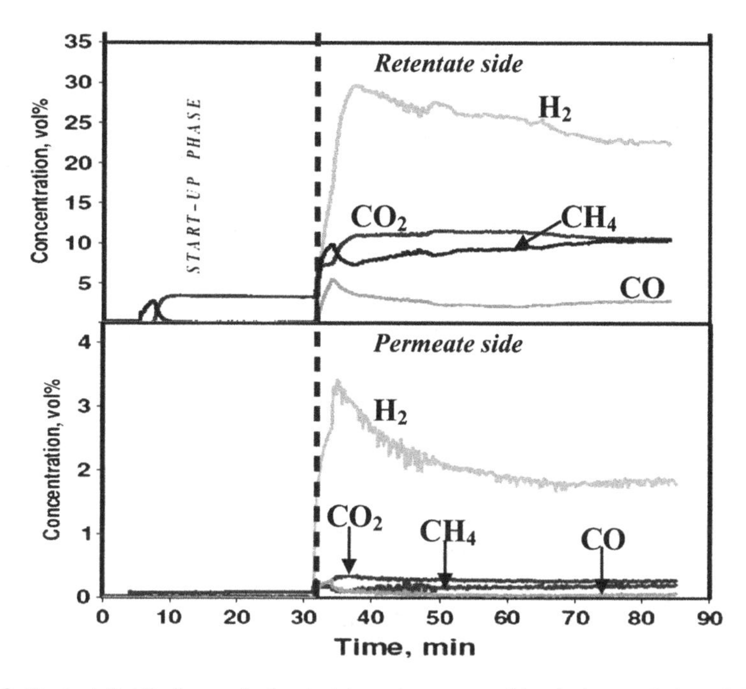

Fig. 6.9 Product distribution on both retentate and permeate sides during catalytic activity tests with INOCERMIC membrane at 1 atm

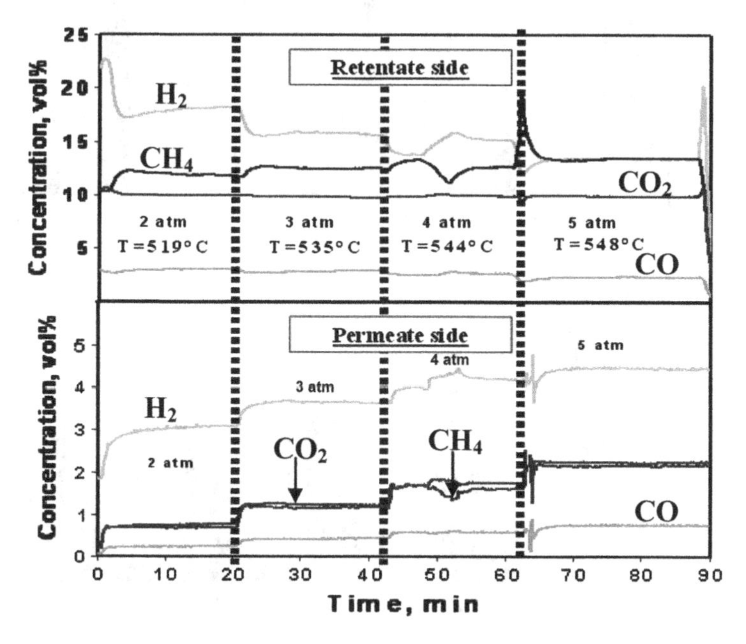

Fig. 6.10 Temperature and product distribution on both retentate and permeate sides during catalytic activity tests with INOCERMIC membrane by changing the pressure from 2 to 5 atm

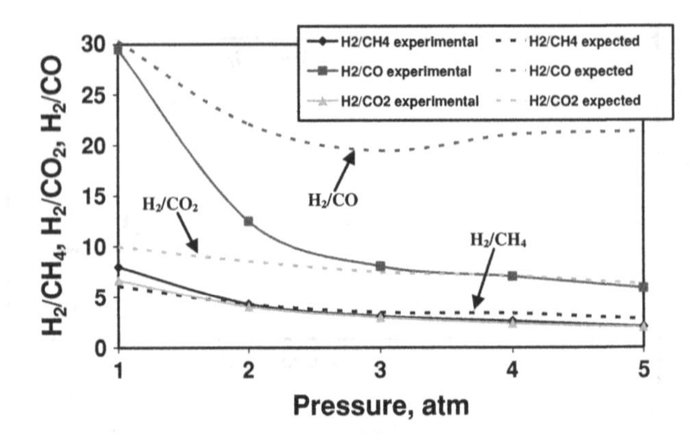

Fig. 6.11 Experimental and expected H_2/CH_4, H_2/CO and H_2/CO_2 ratios for Knudsen diffusion with the INOCERMIC membrane

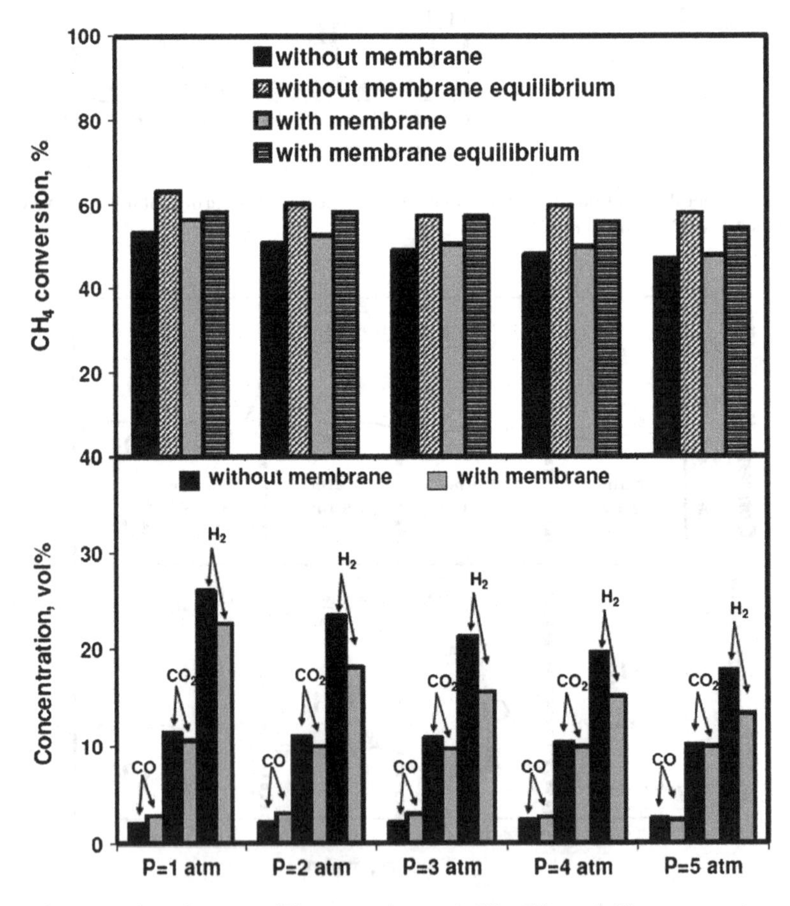

Fig. 6.12 Comparison between CH_4 conversion and CO, CO_2 and H_2 concentration at the reactor outlet in the presence and absence of the membrane

mechanism due to the molecular size must be taken into account. Indeed the enhanced transport of CO and CO_2 through the alumina-based membrane can be explained with their lower molecular size with respect to that of the CH_4 molecule (CO_2 = 0.33 nm, CO = 0.376 nm, CH_4 = 0.38 nm) [84, 85]. Moreover, an increase of pressure leads to a decrease of all separation factors, indeed at high pressures the contribution of transport through membrane cracks could be enhanced [86].

For a better understanding of the effect of the integration with the ATR reactor, the comparison between CH_4 conversion and CO, CO_2 and H_2 concentration at the reactor outlet in the presence and absence of the membrane is reported in Fig. 6.12. For CH_4 conversion, thermodynamic calculations carried out at the temperature inside the catalytic bed are also reported.

It can be observed that at each operating pressure investigated the presence of the membrane allows to reach a higher CH_4 conversion, closer to the predicted thermodynamic one. Moreover, in the presence of membrane, on the retentate side lower H_2 and CO_2 concentrations and higher CO concentration were observed.

6.4 Future Perspectives

The experimental feasibility of the integration of a membrane inside an ATR fixed bed reactor has been demonstrated. High attention must be paid to both catalyst formulation and arrangement/integration with membrane in order to guarantee its performance and stability. Clearly, even if the presence of the membrane could favour the achievement of higher CH_4 conversion at lower temperature, the peculiar temperature profile established along the catalytic bed in an ATR reactor is not very suitable for the integration of the membrane. Accordingly, the reactor architecture is more complex if compared with that in the absence of the membrane, but the advantage observed in terms of fuel processor volume reduction, due to water gas shift and preferential oxidation stages elimination, may drive towards an optimization and diffusion of this innovative solution.

The membrane reactor technology is not yet well established and some deficiencies have to be overcome before its implementation at larger scale. In particular, future research should be devoted to the preparation of membranes able to work for a long period of time at high temperature and pressures and even more resistant in aggressive environments. Furthermore, the sealing of membranes into modules is also another point to be improved in order to avoid problems of streams mixing or bypassing during the reaction tests. Finally, the optimization of engineering in module preparation will be a crucial step for increasing the membrane reactor efficiency. All these optimizations, when carried out, will provide for a technology standardization that will enable for its diffusion. By this way, even if at the moment the costs linked to this solution are quite high (in particular, those relevant to membrane fabrication), it should be possible that the establishment of an economy of scale will enable for a costs depreciation, making this solution more competitive with those currently proposed (refer to Chap. 3).

References

1. Song X, Guo Z (2006) Technologies for direct production of flexible H_2/CO synthesis gas. Energy Convers Manag 47:560–569
2. Navarro RM, Peña MA, Fierro JLG (2007) Hydrogen production reactions from carbon feedstocks: fossil fuels and biomass. Chem Rev 107:3952–3991
3. Holladay JD, Hu J, King DL, Wang Y (2009) An overview of hydrogen production technologies. Cat Today 139:244–260
4. Haldor Topsoe A/S (1988) Hydrocarbon Process 67:77
5. Aasberg-Petersen A, Bak Hansen J-H, Christensen TS, Dybkjaer I, Christensen PS, Stub Nielsen C, Winter Madsen SEL, Rostrup-Nielsen JR (2001) Technologies for large-scale gas conversion. Appl Catal A 221:379–387
6. Joensen F, Rostrup-Nielsen JR (2002) Conversion of hydrocarbons and alcohols for fuel cells. J Power Sources 105:195–201
7. Heinzel A, Vogel B, Hübner P (2002) Reforming of natural gas-hydrogen generation for small scale stationary fuel cell systems. J Power Sources 105:202–207
8. Vermeiren WJM, Blomsma E, Jacobs PA (1992) Catalytic and thermodynamic approach of the oxyreforming reaction of methane. Catal Today 13:427–436
9. Lee SHD, Applegate DV, Ahmed S, Calderone SG, Harvey TL (2005) Hydrogen from natural gas. Part I: autothermal reforming in an integrated fuel processor. Int J Hydrogen Energy 30:829–842
10. Krumpelt M, Krause TR, Carter JD, Kopasz JP, Ahmed S (2002) Fuel processing for fuel cell systems in transportation and portable power applications. Cat Today 77:3–16
11. Wang HM (2008) Experimental studies on hydrogen generation by methane autothermal reforming over nickel-based catalyst. J Power Sources 177:506–511
12. Hoang DL, Chan SH, Ding OL (2006) Hydrogen production for fuel cells by autothermal reforming of methane over sulfide nickel catalyst on a gamma alumina support. J Power Sources 159:1248–1257
13. Horn R, Williams KA, Degenstein NJ, Bitsch-Larsen A, Dalle Nogare D, Tupy SA, Schmidt LD (2007) Methane catalytic partial oxidation on autothermal Rh and Pt foam catalysts: oxidation and reforming zones, transport effects, and approach to thermodynamic equilibrium. J Catal 249:380–393
14. Hoang DL, Chan SH (2004) Modeling of a catalytic autothermal methane reformer for fuel cell applications. Appl Catal A 268:207–216
15. Ding OL, Chan SH (2008) Autothermal reforming of methane gas—modelling and experimental validation. Int J Hydrogen Energy 33:633–643
16. Faur Ghenciu A (2002) Review of fuel processing catalysts for hydrogen production in PEM fuel cell systems. Curr Opin Solid State Mater Sci 6:389–399
17. Takeguchi T, Furukawa S-N, Inoue M, Eguchi K (2003) Autothermal reforming of methane over Ni catalysts supported over CaO–CeO_2–ZrO_2 solid solution. Appl Catal A 240:223–233
18. Choudhary VR, Mondal KC, Mamman AS (2005) High-temperature stable and highly active/selective supported $NiCoMgCeO_x$ catalyst suitable for autothermal reforming of methane to syngas. J Catal 233:36–40
19. Cai X, Dong X, Lin W (2006) Autothermal reforming of methane over Ni catalysts supported on CuO–ZrO_2–CeO_2–$A1_2O_3$. J Nat Gas Chem 15:122–126
20. Dong X, Cai X, Song Y, Lin W (2007) Effect of transition metals (Cu, Co and Fe) on the autothermal reforming of methane over Ni/$Ce_{0.2}Zr_{0.1}Al_{0.7}O_\delta$ catalyst. J Nat Gas Chem 16:31–36

21. Cai X, Cai Y, Lin W (2008) Autothermal reforming of methane over Ni catalysts supported over ZrO_2–CeO_2–Al_2O_3. J Nat Gas Chem 17:201–207
22. Chen X, Tadd AR, Schwank JW (2007) Carbon deposited on Ni/Ce–Zr–O isooctane autothermal reforming catalysts. J Catal 251:374–387
23. Villegas L, Masset F, Guilhaume N (2007) Wet impregnation of alumina-washcoated monoliths: effect of the drying procedure on Ni distribution and on autothermal reforming activity. Appl Catal A 320:43–55
24. Souza MMVM, Schmal M (2005) Autothermal reforming of methane over Pt/ZrO_2/Al_2O_3 catalysts. Appl Catal A 281:19–24
25. Ruiz JAC, Passos FB, Bueno JMC, Souza-Aguiar EF, Mattos LV, Noronha FB (2008) Syngas production by autothermal reforming of methane on supported platinum catalysts. Appl Catal A 334:259–267
26. Li B, Maruyama K, Nurunnabi M, Kunimori K, Tomishige K (2004) Temperature profiles of alumina-supported noble metal catalysts in autothermal reforming of methane. Appl Catal A 275:157–172
27. Kolb G, Baier T, Schürer J, Tiemann D, Ziogas A, Ehwald H, Alphonse P (2008) A micro-structured 5 kW complete fuel processor for iso-octane as hydrogen supply system for mobile auxiliary power units. Part I: development of autothermal reforming catalyst and reactor. Chem Eng J 137:653–663
28. Qi A, Wang S, Ni C, Wu D (2007) Autothermal reforming of gasoline on Rh-based monolithic catalysts. Int J Hydrogen Energy 32:981–991
29. Kaila RK, Gutiérrez A, Korhonen ST, Krause AOI (2007) Autothermal reforming of n-dodecane, toluene, and their mixture on mono- and bimetallic noble metal zirconia catalysts. Catal Lett 115:70–78
30. Qi A, Wang S, Fu G, Wu D (2005) Autothermal reforming of n-octane on Ru-based catalysts. Appl Catal A 293:71
31. Kaila RK, Krause AOI (2006) Autothermal reforming of simulated gasoline and diesel fuels. Int J Hydrogen Energy 31:1934–1941
32. Kaila RK, Gutiérrez A, Krause AOI (2008) Autothermal reforming of simulated and commercial diesel: The performance of zirconia-supported RhPt catalyst in the presence of sulphur. Appl Catal B 84:324–331
33. Cheekatamarla PK, Lane AM (2005) Catalytic autothermal reforming of diesel fuel for hydrogen generation in fuel cells I. Activity tests and sulfur poisoning. J Power Sources 152:256–263
34. Cheekatamarla PK, Lane AM (2006) Catalytic autothermal reforming of diesel fuel for hydrogen generation in fuel cells II. Catalyst poisoning and characterization studies. J Power Sources 154:223–231
35. Shamsi A, Baltrus JP, Spivey JJ (2005) Characterization of coke deposited on Pt/alumina catalyst during reforming of liquid hydrocarbons. Appl Catal A 293:145–152
36. Recupero V, Pino L, Vita A, Cipitì F, Cordaro M, Laganà M (2005) Development of a LPG fuel processor for PEFC systems: laboratory scale evaluation of autothermal reforming and preferential oxidation subunits. Int J Hydrogen Energy 30:963–971
37. Ferrandon M, Krause T (2006) Role of the oxide support on the performance of Rh catalysts for the autothermal reforming of gasoline and gasoline surrogates to hydrogen. Appl Catal A 311:135–145
38. Yuan Z, Ni C, Zhang C, Gao D, Wang S, Xie Y, Okada A (2009) Rh/MgO/$Ce_{0.5}Zr_{0.5}O_2$ supported catalyst for autothermal reforming of methane: the effects of ceria-zirconia doping. Catal Today 146:124–131
39. Cao L, Pan L, Ni C, Yuan Z, Wang S (2010) Autothermal reforming of methane over Rh/$Ce_{0.5}Zr_{0.5}O_2$ catalyst: effects of the crystal structure of the supports. Fuel Process Technol 91:306–312
40. Cheekatamarla PK, Lane AM (2005) Efficient bimetallic catalysts for hydrogen generation from diesel fuel. Int J Hydrogen Energy 30:1277–1285

41. Cheekatamarla PK, Lane AM (2006) Efficient sulfur-tolerant bimetallic catalysts for hydrogen generation from diesel fuel. J Power Sources 153:157–164
42. Dias JAC, Assaf JM (2004) Autothermal reforming of methane over Ni/γ-Al_2O_3 catalysts: the enhancement effect of small quantities of noble metals. J Power Sources 130:106–110
43. Dias JAC, Assaf JM (2005) Autoreduction of promoted Ni/γ-Al_2O_3 during autothermal reforming of methane. J Power Sources 139:176–181
44. Dias JAC, Assaf JM (2008) Autothermal reforming of methane over Ni/γ-Al_2O_3 promoted with Pd. The effect of the Pd source in activity, temperature profile of reactor and in ignition. Appl Catal A 334:243–250
45. Gökaliler F, Selen Çağlayan B, İlsen Önsan Z, Erhan Aksoylu A (2008) Hydrogen production by autothermal reforming of LPG for PEM fuel cell applications. Int J Hydrogen Energy 33:1383–1391
46. Parizotto NV, Zanchet D, Rocha KO, Marques CMP, Bueno JMC (2009) The effect of Pt promotion on the oxi-reduction properties of alumina supported nickel catalysts for oxidative steam-reforming of methane: temperature-resolved XAFS analysis. Appl Catal A 366:122–129
47. Dantas SC, Escritori JC, Soares RR, Hori CE (2010) Effect of different promoters on Ni/$CeZrO_2$ catalyst for autothermal reforming and partial oxidation of methane. Chem Eng J 156:380–387
48. Liu D-J, Krumpelt M, Chien H-T, Sheen S-H (2006) Critical issues in catalytic diesel reforming for solid oxide fuel cells. J Mater Eng Perform 15:442–444
49. Liu D-J, Krumpelt M (2005) Activity and structure of perovskites as diesel-reforming catalysts for solid oxide fuel cell. Int J Appl Ceram Technol 2:301–307
50. Qi A, Wang S, Fu G, Ni C, Wu D (2005) La–Ce–Ni–O monolithic perovskite catalysts potential for gasoline autothermal reforming system. Appl Catal A 281:233–246
51. Erri P, Dinka P, Varma A (2006) Novel perovskite-based catalysts for autothermal JP-8 fuel reforming. Chem Eng Sci 61:5328–5333
52. Mawdsley JR, Krause TR (2008) Rare earth-first-row transition metal perovskites as catalysts for the autothermal reforming of hydrocarbon fuels to generate hydrogen. Appl Catal A 334:311–320
53. Rostrup-Nielsen JR (2000) New aspects of syngas production and use. Catal Today 63:159
54. Ahmed S, Krumpelt M (2001) Hydrogen from hydrocarbon fuels for fuel cells. Int J Hydrogen Energy 26:291–301
55. Maestri M, Beretta A, Groppi G, Tronconi E, Forzatti P (2005) Comparison among structured and packed-bed reactors for the catalytic partial oxidation of CH_4 at short contact times. Catal Today 105:709–717
56. Giroux T, Hwang S, Liu Y, Ruettinger W, Shore L (2005) Monolithic structures as alternatives to particulate catalysts for the reforming of hydrocarbons for hydrogen generation. Appl Catal B 55:185–200
57. Kikuchi E (2000) Membrane reactor application to hydrogen production. Catal Today 56:97–101
58. Ciambelli P, Palma V, Palo E, Sannino D (2005) Hydrogen production via catalytic autothermal reforming of methane. In: Proceedings of "7th World Congress of Chemical Engineering", Glasgow (Scotland) July 10–14, p 225
59. Palo E (2007) Structured catalysts for hydrogen production by methane autothermal reforming. Ph.D Thesis, University of Salerno
60. Ciambelli P, Palma V, Palo E, Iaquaniello G (2009) Natural gas autothermal reforming: an effective option for a sustainable distributed production of hydrogen. In: Barbaro P, Bianchini C (eds) Catalysis for sustainable energy production. Wiley, Weinheim, pp 287–319
61. Iaquaniello G, Mangiapane A, Ciambelli P, Palma V, Palo E (2005) Small scale hydrogen production. Chem Eng Trans 8:19–26

62. Ciambelli P, Palma V, Palo E, Iaquaniello G, Mangiapane A, Cavallero P (2007) Energy sustainable development through methane autothermal reforming for hydrogen production. AIDIC Conf Series 8:67–76
63. Ciambelli P, Palma V, Palo E (2008) Comparison of ceramic honeycomb monolith and foam as Ni catalyst carrier for methane autothermal reforming. Catal Today. Catal Today 155:92–100
64. Ciambelli P, Palma V, Palo E, Villa P (2008) Reattore catalitico autotermico con profilo di temperatura piatto per la produzione di idrogeno da idrocarburi leggeri. Italian Patent Pending SA2008A/000023
65. Ciambelli P, Palma V, Palo E, Villa P (2010) Autothermal catalytic reactor with flat temperature profile. PCT Int. Appl. WO 2010/016027
66. Ciambelli P, Palma V, Palo E, Iaquaniello G (2009) Experimental and economical approach to the integration of a kW-scale CH_4-ATR reactor with a WGS stage. Chem Eng Trans 18:499–504
67. Palma V, Palo E, Ciambelli P (2009) Structured catalytic substrates with radial configurations for the intensification of the WGS stage in H_2 production. Catal Today 147:S107–112
68. Lattner JR, Harold MP (2004) Comparison of conventional and membrane reactor fuel processors for hydrocarbon-based PEM fuel cell systems. Int J Hydrogen Energy 29:393–412
69. Lattner JR, Harold MP (2005) Comparison of methanol-based fuel processors for PEM fuel cell systems. Appl Catal B 56:149–196
70. Tiemersma TP, Patil CS, van Sint Annaland M, Kuipers JAM (2006) Modelling of packed bed membrane reactors for autothermal production of ultrapure hydrogen. Chem Eng Sci 61:1602–1616
71. Feng W, Tan T, Ji P, Zheng D (2006) Exploration of hydrogen production in a membrane reformer. AIChE J 52:2260–2270
72. Feng W, Ji P (2007) Multistage two-membrane ATR reactors to improve pure hydrogen production. Chem Eng Sci 62:6349–6360
73. Hüppmeier J, Baune M, Thöming J (2008) Interactions between reaction kinetics in ATR-reactors and transport mechanism in functional ceramic membranes: a simulation approach. Chem Eng J 142:225–238
74. Chen Z, Yan Y, Elnashaie SSEH (2003) Modeling and optimization of a novel membrane reformer for higher hydrocarbons. AIChE J 49:1250–1265
75. Prasad P, Elnashaie SSEH (2003) Coupled steam and oxidative reforming for hydrogen production in a novel membrane circulating fluidized-bed reformer. Ind Eng Chem Res 42:4715–4722
76. Chen Z, Yan Y, Elnashaie SSEH (2003) Novel circulating fast fluidized-bed membrane reformer for efficient production of hydrogen from steam reforming of methane. Chem Eng Sci 58:4335–4349
77. Patil CS, van Sint Annaland M, Kuipers JAM (2005) Design of a novel autothermal membrane-assisted fluidized-bed reactor for the production of ultrapure hydrogen from methane. Ind Eng Chem Res 44:9502–9512
78. Chen Z, Grace JR, Lim CJ, Li A (2007) Experimental studies of pure hydrogen production in a commercialized fluidized-bed membrane reactor with SMR and ATR catalysts. Int J Hydrogen Energy 32:2359–2366
79. Mahecha-Botero A, Boyd T, Gulamhusein A, Comyn N, Lim CJ, Grace JR, Shirasaki Y, Yasuda I (2008) Pure hydrogen generation in a fluidized-bed membrane reactor: experimental findings. Chem Eng Sci 63:2752–2762
80. Gallucci F, Van Sint Annaland M, Kuipers JAM (2008) Autothermal reforming of methane with integrated CO_2 capture in a novel fluidized bed membrane reactor. Part 1: experimental demonstration. Top Catal 51:133–145
81. Gallucci F, Van Sint Annaland M, Kuipers JAM (2008) Autothermal reforming of methane with integrated CO_2 capture in a novel fluidized bed membrane reactor. Part 2: comparison of reactor configurations. Top Catal 51:146–157

82. Simakov DSA, Sheintuch M (2008) Design of a thermally balanced membrane reformer for hydrogen production. AIChE J 54:2735–2750
83. Lu GQ, Diniz da Costa JC, Duke M, Giessler S, Socolow R, Williams RH, Kreutz T (2007) Inorganic membranes for hydrogen production and purification: a critical review and perspective. J Colloid Interface Sci 314:589–603
84. Breck DW (1974) Zeolite molecular sieves: structure, chemistry and use. Wiley, New York, p 636
85. Lee D, Oyama ST (2002) Gas permeation characteristics of a hydrogen selective supported silica membrane. J Membr Sci 210:291–306
86. de Lange RSA, Keizer K, Burggraaf AJ (1995) Analysis and theory of gas transport in microporous sol–gel derived ceramic membranes. J Membr Sci 104:81–100

Chapter 7
Technical and Economical Evaluation of WGSR

Paolo Ciambelli, Vincenzo Palma, Emma Palo, Jan Galuszka, Terry Giddings and Gaetano Iaquaniello

7.1 Process Description

The water–gas shift reaction (WGSR) (7.1) was discovered more than two centuries ago by Italian physicist Felice Fontana in 1780 [1].

$$CO + H_2O \leftrightarrow CO_2 + H_2 \tag{7.1}$$

Although the first report was published in 1888 by Mond and Longer [2], the technical importance of the WGSR was not recognized until the development of the Haber process. Currently, the WGSR is used in various chemical processes, such as hydrogen and ammonia production, Fisher–Tropsch and methanol synthesis. Also, it is considered to be an important process for the removal of CO in small-scale future power generation, based on fuel cells for both mobile and stationary applications.

The WGSR is not affected by pressure and is moderately exothermic ($\Delta H^0_{298\ K} = -41.1$ kJ mol^{-1}); therefore, its equilibrium constant decreases with temperature, favoring higher CO conversions at lower temperatures. Addition of

P. Ciambelli (✉) and V. Palma
Department of Chemical and Food Engineering, University of Salerno, 84084 Fisciano, SA, Italy
e-mail: pciambelli@unisa.it

J. Galuszka and T. Giddings
Natural Resources Canada, CanmetENERGY, 1 Haanel Drive, Ottawa, ON K1A 1M1, Canada
e-mail: galuszka@NRCan.gc.ca

G. Iaquaniello and E. Palo
Tecnimont KT, S.p.A, Viale Castello della Magliana 75, 00148 Rome, Italy
e-mail: Iaquaniello.G@tecnimontkt.it

M. De Falco et al. (eds.), *Membrane Reactors for Hydrogen Production Processes*, DOI: 10.1007/978-0-85729-151-6_7, © Springer-Verlag London Limited 2011

greater than stoichiometric quantities of steam improves the conversion. Since under adiabatic conditions, the heat of reaction increases process temperature, conversion in a single catalyst bed is thermodynamically limited. A significant improvement is obtained with a double catalyst bed operation, with the second bed operating at the lowest possible inlet temperature. Usually, the high temperature shift (HTS) reactor operates in the range of 350–420°C with an iron-chromium oxide-based catalyst, and the low-temperature shift (LTS) reactor operates in the range of 180–340°C with a Cu–ZnO/Al_2O_3 catalyst [3]. With this reactor architecture, the exit concentration of CO could be as low as 0.1–0.3%. However, some serious disadvantages could also be identified. The HTS catalyst has low activity at lower temperatures, and the process is thermodynamically limited at high temperatures. The low temperature Cu–ZnO catalyst is sensitive to air exposure, promotes temperature excursions and requires lengthy preconditioning for intermittent operation. Thus, a significant effort to improve the performance of iron [4–10] and Cu–ZnO [11–13]-based catalysts by optimizing the catalyst preparation and formulation is evident in the literature. Because of its unique redox properties, ceria was tested as CuO-based catalyst support [14–18]. Also, research attention has been recently focused on the use of noble metals such as Pt [19–27] and Au [28–34].

One of the potential applications of the WGSR is the integrated gasification combined (IGCC) process, in which the gasification of coal produces synthesis gas and the WGSR converts the CO to produce additional H_2. Despite coal being an available raw material with relatively stable cost for H_2 production as an alternative to gaseous and liquid hydrocarbons, the IGCC process presents additional challenges because of the lower pressure and lower H_2 content of the clean gas.

In the nineties, the first generation of IGCC plants appeared, and although they are reliable and have shown environmental benefits, further improvements to simplify the process, increase efficiency and lower costs are needed to advance the commercial outlook of the IGCC scheme. Despite the existence of several commercial, entrained flow gasification systems for the production of fuel gas or syngas, the process has yet to be demonstrated on the commercial scale as part of an integrated plant for the production of H_2 with the collection and storage of species of environmental concern such as CO_2. These systems, including Shell, Texaco, Destec and Prenflo gasifiers, are similar in that they utilize pulverized coal and operate at pressures in the range of 20–70 bar and temperatures around 1,500°C with very high fuel heating rates. Differences in these systems include the way in which the fuel is introduced, the concentration of steam and methods employed for heat recovery.

A block flow diagram for a conventional IGCC plant is shown in Fig. 7.1. The hot syngas from the gasifier is passed through a Heat Recovery Steam Generation Unit, where it is cooled to below 300°C. Next, the syngas is passed to the Filtration Unit, where fine particulates are removed. Further clean-up is carried out in the Water Wash Unit, followed by sulphur removal and recovery. The process of separating sulphur compounds consists of a catalytic hydrolysis reactor in which COS is transformed into H_2S and HCN in NH_3. The H_2S is recovered by MDEA in an absorption column. The MDEA is regenerated in a second column where H_2S is

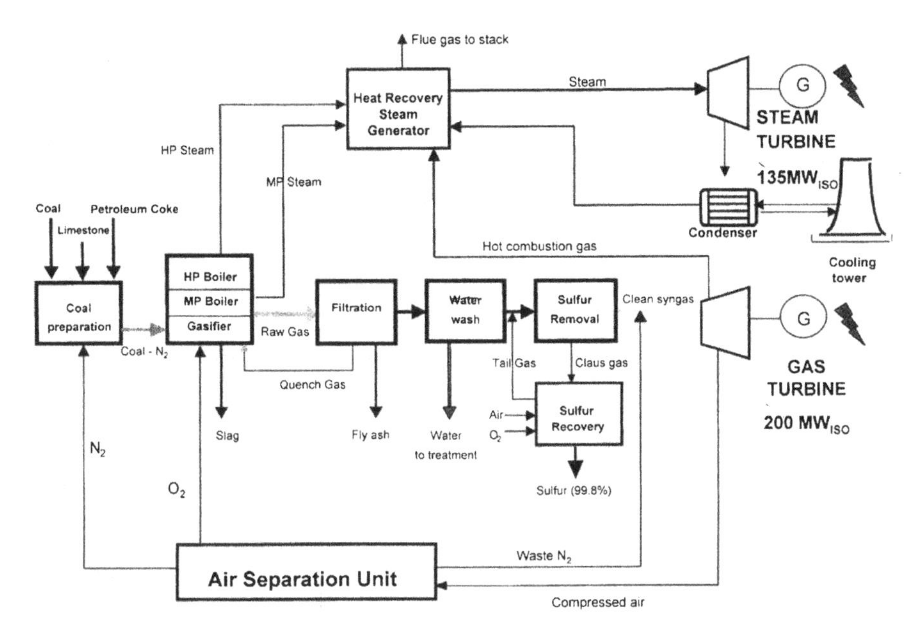

Fig. 7.1 IGCC schematic block diagram

separated and sent to the Claus recovery unit, where it is converted to sulphur and water. For the Puertollano IGCC plant, the typical composition of the clean gas is within the following range [35]: H_2 19.3–24.87%, Ar 0.76–1.33%, CO 57.67–63.37%, CO_2 1.51–3.72%, COS 0.14–20 ppmv, H_2S 0–10 ppmv, CH_4 35–120 ppmv, CNH 1–3 ppmv, NH_3 3 ppmv, SO_2 0–9 ppmv and CS_2 0–5 ppmv. This stream is collected at a temperature of 126.4–144.6°C and a pressure of 18.8–21.2 barg. The clean gas is then sent to a gas turbine where a combined cycle is applied for electricity generation.

A schematic of the conventional WGSR is shown in Fig. 7.2. For the conventional coal-to-hydrogen process, the WGSR takes place after the filtration unit and before the acid gas removal unit. For this reason, a sulphur tolerant catalyst must be used for the WGSR.

Steam is added to the syngas before entering the WGSR reactor. The WGSR is assumed to reach equilibrium by the exit of each reactor, which is estimated to give a CO conversion of at least 80%, depending on the adiabatic temperature of the reactor. All of the H_2S and most of the CO_2 are removed in the acid gas removal unit. Available acid gas removal technologies include low temperature absorption by amines, glycol and chilled methanol; hot potassium carbonate process; and separation by low temperature polymeric membranes. All the low temperature processes require gas cooling and heat recovery leading to energy losses. Also, significant energy is required for regeneration in solvent- and reagent-based systems. Finally, hydrogen is purified by PSA and used for power generation. The PSA tail gas, which still contains CH_4, H_2 and CO, is sent to a boiler for steam generation.

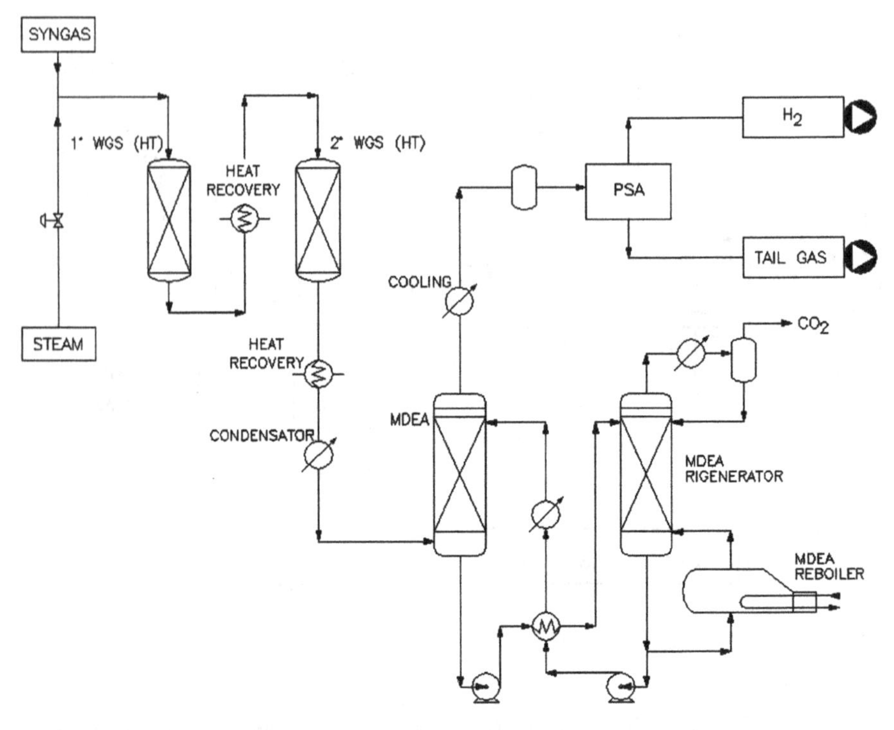

Fig. 7.2 Conventional process for shift conversion

From what is described, it is quite clear that a WGSR is required in the process scheme only when CO_2 removal or further production of H_2 is necessary. The existing technologies using absorption with physical solvents, namely Rectisol and Selexol, or chemicals such as m-DEA, can be readily applied to capture CO_2 from the coal gas at IGCC plants. However, such applications have significant energy and equipment requirements for solvent circulation and heat exchange, which results in a reduction in efficiency by ca. 6% (i.e. from ca. 42 to 36%). Consequently, new systems need to be developed, such as advanced membranes or adsorbents that offer significant potential for efficiency improvements and cost reduction for CO_2 capture in gasification. Cheaper, more efficient H_2 separation membranes as passive separators are promising alternatives to conventional scrubbers. Further, the integration of membranes with WGS reactors is promising since, in addition to H_2 separation and CO_2 recovery, selective removal of H_2 from the reactor allows for further conversion of CO beyond that possible through conventional WGSRs. Several studies available in the open literature report the feasibility of the use of WGSR membrane reactors in IGCC systems [36–42].

As a reference case, the simulation of a conventional WGSR unit was carried out. A good prediction of WGSR effluent stream was obtained using a temperature approach based on the ShiftMax 120 catalyst produced by Sud-Chemie [43] assuming WGSR feed gas compositions that are typically produced by the

Table 7.1 Material balances on the two HTS reactors

Compounds	I HTS IN (kmol/h)	I HTS OUT (kmol/h)	II HTS OUT (kmol/h)
CO	974.2	181.60	87.71
H_2	355.5	1148.1	1242.0
CO_2	62.3	854.9	948.8
N_2	201.3	201.3	201.3
Ar	16.74	16.70	16.74
H_2O	3,220	2,427	2,333.5

Puertolano gasification plant. For the high CO concentrations in the syngas, it is necessary to use two HTS reactor stages, each of which operate with an inlet temperature of 350°C.

The steam fed to the first HTS reactor from the gasifier has a steam to CO ratio of about 2. Extra steam from 'battery limits' is required to bring the steam-to-CO ratio to 3.3 to meet the requirements of the catalyst. In Table 7.1 the material balance for the two HTS reactors is shown, based on hydrogen production of 25,000 Nm^3/h.

The CO conversion in the first HTS reactor is 81.4%. Since the WGSR is exothermic, heat is removed after the first HTS reactor to render the thermodynamics favourable for further conversion of CO. Of the CO remaining from the first HTS reactor, 51.7% is converted in the second HTS reactor to give an overall plant CO conversion of 91%. The hydrogen from the WGSR is recovered by PSA. The PSA off-gas contains small amounts of H_2 and CO that can be burned in a combustor. In order to increase CO conversion, a recycle of this gas may also be considered. Because of the high percentage of N_2 and CO_2 in the off-gas, this option does not result in significant improvements in plant efficiency because of the increased plant size requirements. Therefore, this off-gas is considered useful for heat recovery only by its LHV (1,141 kcal/kg) and flow rate. A H_2 recovery of 85% has been assumed for the PSA unit.

7.2 Membrane-Assisted IGCC

It is considered that combining the IGCC and hydrogen membrane reactor technologies could radically improve a commercial outlook for the IGCC scheme application to power production with zero CO_2 emissions, leading to elimination of coal-fired power plants [36, 44, 45]. The expected significant process simplification and intensification capitalizes on a paradigm offered by membrane reactors, which allows for the reaction and separation to be combined in one step. An example of the expected gain in CO conversion by hydrogen removal during the WGSR is shown in Fig. 7.3.

Membrane integration with the IGCC could practically be realized in two possible reactor architectures determined by a relative placement of the membranes and catalysts. In the simplest arrangement, two membrane passive separators could be located in front of each of the conventional HTS and LTS

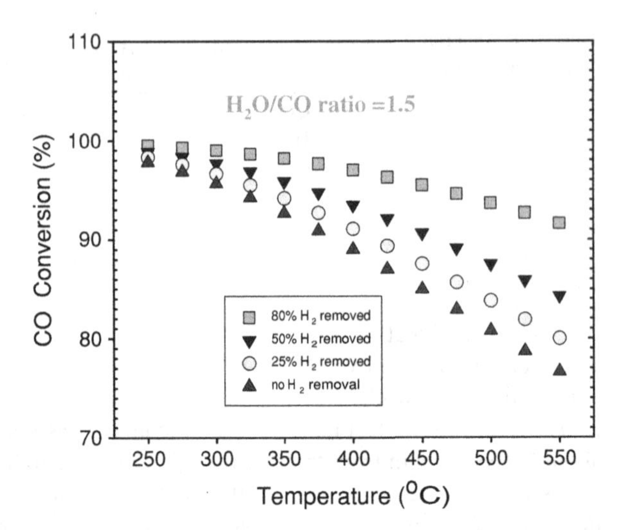

Fig. 7.3 CO conversion increase during WGSR stimulated by hydrogen removal

reactors, forming the so-called open architecture (OA). In this configuration, carbon capture would consist of a CO_2 removal unit and final purification with PSA. The more complex but less process-intensive closed architecture (CA) would eliminate the conventional HTS and LTS reactors entirely, replacing them by a single membrane reactor containing suitable WGS catalyst and thus fully capitalizing on the principle of reaction and separation in one step. Hydrogen would be removed through the membrane as it is produced, circumventing the thermodynamic limitation and producing a clean, sequestration-ready CO_2 stream. This arrangement may currently be considered technically too complex from an engineering or maintenance point of view, but it seems to be the most studied case on the laboratory scale. Nevertheless, the OA should not be considered an alternative to CA, but rather the first simpler step towards commercial realization of hydrogen selective membrane technology integration with the WGSR process. Both these reactor architectures are reviewed in this chapter.

7.3 Closed Architecture

A growing interest in the membrane-assisted WGSR is manifested by a substantial volume of the published pertinent literature that could be grouped around the two most popular hydrogen permselective membranes, based either on palladium [46–55] or silica [56–59]. The cited references cover only the most recent and most representative examples.

Basile et al. [46] studied the WGSR in a palladium membrane reactor and showed the importance of the membrane preparation method in obtaining high-quality membrane materials. Magnetron sputtering, physical vapour deposition and co-condensation techniques were applied to realize submicron palladium membranes. The best membrane increased CO conversion up to 99.89% at about 330°C and stable performance was reported for more than 2 months [47].

Iyoha et al. [50] assessed the performance of a Pd and $Pd_{80\ wt\%}Cu$ membrane reactor at 900°C with a 241 kPa trans-membrane pressure differential intended to be positioned downstream of a coal gasifier. No catalyst was used as it was expected that the membrane tubes would sufficiently catalyse and further enhance the fast rate of the WGSR at this temperature. There was a significant increase in CO conversion from the equilibrium value of 54–93% with the Pd membrane, while conversion of 66% was attained with the $Pd_{80\ wt\%}Cu$ membrane. The markedly lower conversion with the $Pd_{80\ wt\%}Cu$ membrane was attributed to its lower H_2 permeance. However, after about 8 days on stream, pinhole formation was confirmed by SEM-EDS for both membranes. In follow-up WGSR studies [51], these membranes were applied to the simulated coal gasification syngas feeds. CO conversion of 99.7% was achieved at 900°C in a counter-current, Pd multi-tube membrane reactor operated at a 2-s residence time. The conversion of CO was considerably higher than the approximately 32% equilibrium conversion allowed in a conventional reactor. As found previously, the $Pd_{80\ wt\%}Cu$ membrane tubes gave a significantly lower CO conversion of only 68%. Exposure of both membranes to syngas mixtures containing H_2S at levels below the threshold required for the formation of thermodynamically stable sulphides ($H_2S/H_2 < \sim 0.0011$) did not affect the mechanical strength of the membranes, but caused a steep drop in CO conversion because of deactivation of the catalyst surface. Any further increase in H_2S concentration destroyed the membranes within minutes.

Recently, Barbieri et al. [52] proposed an innovative configuration of a membrane reactor consisting of a conventional fixed bed reactor followed by a membrane reactor loaded with the same catalyst. This configuration was viewed as particularly suitable for reactions characterized by slow kinetics, such as the WGSR, as it stimulates more effective membrane utilization by optimizing the driving force and minimizing the H_2 back permeation to the reaction side. The WGSR was tested between 280 and 320°C using a commercial CuO/CeO_2-based catalyst and Pd–Ag membrane without sweeping for an equimolecular H_2O/CO stream at pressures between 2 and 6 atm and gas hourly space velocity (GHSV) between 2,000 and 10,000 h^{-1}. A significant reduction in the required reaction volume to reach CO conversion similar to that of the conventional reactor was achieved, which should result in a decreased cost of operation.

Bi et al. [53] employed a noble metal base catalyst $Pt/Ce_{0.6}Zr_{0.4}O_2$ and Pd membrane reactor to study the WGSR using feeds obtained by autothermal reforming of natural gas. CO conversion remained above thermodynamic equilibrium up to feed space velocities of 9,100 l kg^{-1} h^{-1} at 350°C, $P_{total} = 1.2$ MPa and $S/C = 3$, but H_2 recovery decreased from 84.8% at space velocity of 4,050 l kg^{-1} h^{-1} to 48.7% at the highest space velocity. This rapid decline of separation performance was attributed to slow H_2 diffusion through the catalyst bed, suggesting that external mass flow resistance has a significant impact on the H_2 permeation rate in such membrane reactors.

Near-complete conversion of CO was obtained by Kikuchi et al. [60] and by Uemiya et al. [61] in the WGSR carried out at 400°C in a double tubular Pd membrane reactor with a commercial iron-chromium oxide as catalyst.

Despite all these studies, a recent economic feasibility study of the membrane-assisted WGSR conducted to quantify the advantage over conventional technology disproved the concept for Pd membrane reactors [62] and supported the use of ceramic membranes [36].

Microporous silica hydrogen permselective membranes have been extensively studied as a potentially more practical alternative to Pd membranes. Very recently, a comprehensive review was published, tackling various aspects of silica membrane synthesis, application and economics [63]. It was made evident that state-of-the-art silica membranes have good hydrogen flux and separation, as well as respectable thermal stability. However, the hydrothermal stability of a silica hydrogen permselective membrane is a key factor in determining its suitability for a commercial application of membrane-assisted processes.

A silica membrane prepared by the soaking–rolling procedure on a porous stainless steel disc [64] was employed by Brunetti et al. [56] for the WGSR from 220 to 290°C and up to 600 kPa using a commercial CuO/CeO_2-based catalyst. The CO conversion always exceeded conversions attainable in a conventional reactor at temperatures higher than 250°C. The CO conversion difference between the conventional and the membrane reactor increased with temperature and was more pronounced at lower reaction pressures. It was determined that the best operating conditions for the membrane reactor were 280°C and 400 kPa.

Application of a silica membrane supported on molecular sieve to the low-temperature WGSR at 280°C was reported by Giessler et al. [57]. Although almost complete conversion of CO was claimed, the low H_2/N_2 separation of the membrane opened up a possibility for CO to cross the membrane to the sweep stream and for the sweep gas to enter the product stream, artificially boosting the conversion.

Battersby et al. [58] investigated the WGSR in silica membrane between 150 and 250°C. The H_2/CO separation increased from 5 to 15 with temperature, but at conversions below 40%, the H_2 driving force for permeation was lower than that for CO, dictating the minimum conversion for which the membrane reactor could be effective. In a follow-up study [59], a cobalt-silica membrane was used to test the WGSR up to 300°C. The initial H_2/CO separation increased with temperature, resulting in up to 95% purity of the permeated H_2. A 7% increase in conversion above that obtained in a conventional reactor was achieved at 300°C and the membrane showed a reasonable on-stream stability for about 200 h. However, the membrane had no effect on conversion below 200°C, indicating that application of a membrane reactor in a kinetically controlled range of the WGSR is perhaps not the best choice of conditions to demonstrate a paradigm offered by membrane reactors, as the equilibrium conversion is already high but the reaction rate is low. The hydrothermal stability of the membrane was most likely an issue at higher temperatures and dictated the choice of these less representative conditions.

The membrane-assisted WGSR was studied by Galuszka et al. [65] using a commercial iron-chromia-based catalyst at 450°C in conventional and silica membrane fixed-bed reactors. The H_2O/CO ratio was kept at 1.75. The catalyst gave a steady performance and up to 80% of the H_2 produced during the WGSR in the membrane reactor was removed through the membrane that had the initial H_2/N_2 separation of $\sim$ 150 and H_2 permeance of $\sim$2.5 (NTP) $cm^3\ cm^{-2}\ min^{-1}\ atm^{-1}$ (0.19×10^{-6} mol $m^{-2}\ s^{-1}\ Pa^{-1}$). A 6–10% increase in CO conversion was achieved in a membrane reactor as compared to a conventional reactor for a simulated feed stream from a gasifier. After about 10 days on-stream, the H_2/N_2 separation of the membrane usually decreased to about 50% of the initial value. After that, further decrease in membrane performance was barely noticeable. The overall excellent performance of the CanmetENERGY membrane evoked great confidence in the demonstrated feasibility of a novel WGS reactor concept and provided a good base for a modelling study.

The kinetics of the WGSR was studied extensively and a great variety of alternative models and different mechanisms were proposed as outlined in [65]. A kinetic expression,

$$r = k_T K_{CO} K_{H_2O} \frac{P_{CO} P_{H_2O}(1-\beta)}{(1 + K_{CO}P_{CO} + K_{H_2O}P_{H_2O} + K_{CO_2}P_{CO_2} + K_{H_2}P_{H_2})^2} \tag{7.2}$$

where

$$\beta = \frac{P_{CO_2} P_{H_2}}{P_{CO} P_{H_2O} K_{eq}} \tag{7.3}$$

imposing a Langmuir–Hinshelwood mechanism [48] to evaluate the reaction rate 'r' of CO consumption, seems to be a popular choice and the modelling of conventional and membrane reactors described by Galuszka et al. [65] was based on this mechanism. The kinetic parameters were first evaluated for the set of experimental data derived from the conventional reactor studies where the three different feeds were used with the same amount of catalyst and variable contact time. Excel Solver program was applied to minimize the sum of squared residuals of the calculated and experimental CO flows.

Then, the evaluated kinetic constants were used to model the membrane reactor data. An example of the results taken from [65] is shown in Fig. 7.4.

The experimental point in Fig. 7.4 was an average integral conversion during a steady-state operation for 30 h on stream run. It was observed [65] that in addition to the experimentally determined H_2 permeance, the water permeance of silica membrane, equal to about 30% of its H_2 permeance, had to be assumed to obtain an adequate fit. Curve A in Fig. 7.4 denoted simply CO conversion based on the amount of hydrogen removed, without considering the kinetics as shown in Fig. 7.3. The calculated conversion of CO along the length of the membrane indicated that only about 12% of the catalyst was needed to bring the system to thermodynamic equilibrium. The rest of the catalyst assisted by membrane provided the additional 6% conversion above the equilibrium level as observed experimentally.

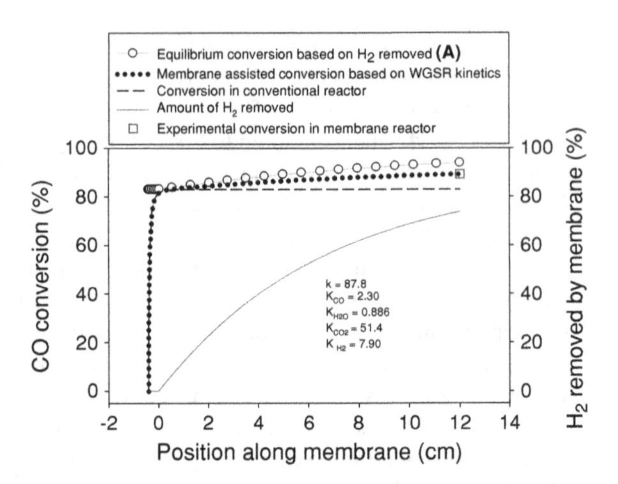

Fig. 7.4 Results of Langmuir–Hinshelwood model applied to membrane reactor-assisted WGSR (reprinted with permission from [65])

These modelling data provided, for the very first time, a solid base enabling a realistic extrapolation of the influence of membrane performance on the extent of the WGSR, including its kinetics. The most interesting conclusion of that study was the estimated real membrane performance that was needed to engage the full length of the catalyst bed and to obtain almost complete conversion of CO. It was calculated that increasing the membrane H_2 permeance by a factor of only two and operating with a H_2O/CO ratio of approximately 3.8 would fully engage the catalyst and increase the CO conversion to about 98% at 450°C. However, hydrogen removal would need to be five times more efficient, as measured by the sweep rate, to keep an adequate driving force for the membrane. Since water permeated through the silica membrane, the increased amount of water was needed to help improve the kinetics of the WGSR and to avoid downstream water depletion in the membrane reactor.

The WGSR in a mixed protonic–electronic conducting $SrCe_{0.9}Eu_{0.1}O_{3-\delta}$ membrane coated on a Ni-$SrCeO_{3-\delta}$ support was studied by Li et al. [66]. At 900°C a 46% increase in CO conversion was achieved, with the feed stream having a H_2O/CO ratio of 2.

Another concept of circumventing the thermodynamic limitation of the WGSR considered in the literature was the application of CO_2 permselective membranes. It was claimed that the use of CO_2-selective membranes may present some advantage compared to H_2-selective membranes because (i) a H_2-rich product is recovered at high pressure (feed gas pressure) and (ii) air can be used to sweep CO_2-rich permeate on the low pressure side of the membrane [67]. A one-dimensional non-isothermal model to simulate the counter-current hollow-fiber membrane reactor was developed by Huang et al. [67]. Based on the modelling study, a CO concentration of less than 10 ppm, a H_2 recovery of greater than 97% and a H_2 concentration of greater than 54% (on the dry basis) are achievable from autothermal reforming syngas. Moreover, if steam reforming syngas is fed into the membrane reactor, then a H_2 concentration greater than 99.6% can be obtained along with a reduced membrane area requirement. The modelling study showed that both the WGSR rate and CO_2 permeation played an important role on overall reactor performance. The model was later verified experimentally by the same authors using a rectangular flat-sheet WGS membrane reactor with an autothermal reforming syngas at 150°C [68]. Further investigations by Huang and Ho [69] considered several parameters such as inlet feed temperature, inlet sweep temperature, feed-side pressure, WGSR catalyst activity and CO inlet feed concentration. An increase in the inlet feed temperature and the inlet sweep temperature (up to about 160°C) resulted in a reduction of the required membrane area. The same result was obtained with an increase in the feed-side pressure.

It needs to noted, however, that the CO_2 permselective membrane, which is in its developmental infancy, is conceptually more difficult than the much more versatile and close to commercialization H_2-selective membrane. The principle of molecular size discrimination, which forms the basis of the hydrogen permselective membrane cannot be applied to CO_2 separation since CO_2 is larger than or of similar molecular size as the species that it must be separated from. The idea of membrane chemical 'functionalization', which is necessary to achieve selective separation of CO_2, is difficult to implement and will most likely result in a significant limitation to the CO_2 membrane range of applications.

7.4 Open Reactor Architecture

The proposed membrane integration with the IGCC in the simplest arrangement of the OA, having membrane passive separators located in front of each conventional WGSR reactor, could be the first step in a practical realization of the membrane-assisted WGSR. A schematic of the OA WGSR reactor is shown in Fig. 7.5. In this scheme the shell and tube membrane separator modules are constructed from silica hydrogen permselective tubular membranes. The performance of a silica membrane used in [65] is shown in Table 7.2.

The membrane separator selectively removes H_2 from the syngas feed so that the conversion of CO in the following WGSR conventional reactor can be more thermodynamically favourable. A hydrogen-rich permeate stream and a medium pressure CO_2-rich stream are obtained as an end product. Because the hydrogen is removed before the WGSR, the volume of the syngas stream is reduced, resulting in a lower steam requirement to maintain appropriate steam concentration to avoid sintering of the catalyst. Owing to the improved thermodynamics resulting from H_2 removal by the membrane, the WGSR may be operated at a higher inlet temperature (515°C) where kinetics is more favourable.

The addition of steam before the second reactor brings the *S/C* ratio to 2.2 and simultaneously lowers the first WGSR reactor outlet stream temperature from 540 to 515°C and prevents catalyst sintering in the second WGSR reactor. The performance of this innovative configuration is reported in Table 7.3.

The CO conversion for the plant predicted for the OA WGSR reactor scheme is 97.5%, as compared to 91% for the reference case of a conventional WGSR, due to

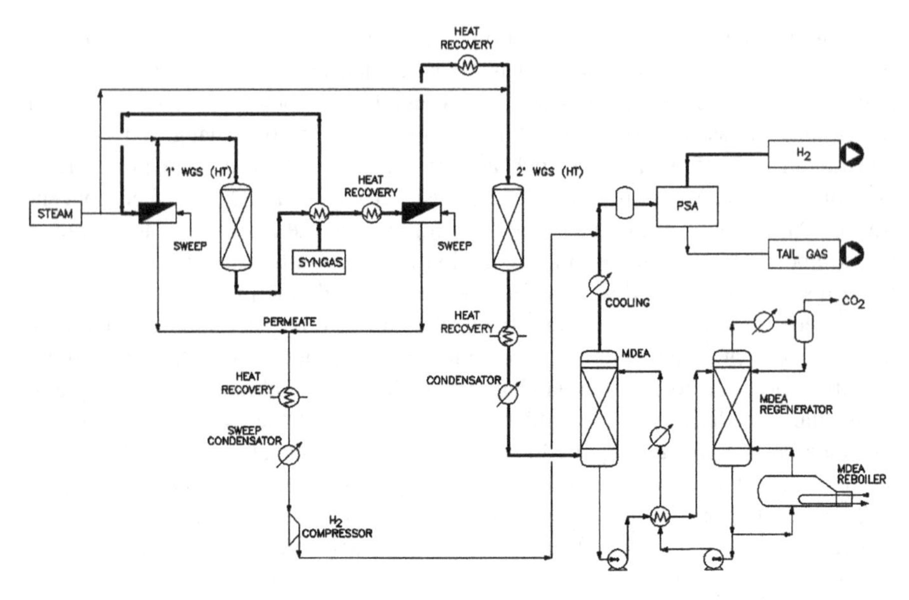

Fig. 7.5 Schematic of OA WGSR reactor

Table 7.2 Silica membrane performance

Temperature (°C)	Permeance (kmol/h m^2 kPa)	Separation (H_2/N_2)
450–500	7.0 E–4	150

Table 7.3 Performance of the OA membrane-based process

Parameter	Value
Feed (kmol/h)	1,449.0
Feed pressure (kPa)	2,000
1st WGS conversion (%)	83.8
2nd WGS conversion (%)	84.9
Plant conversion (%)	97.5
CO_2 absorbed (kmol/h)	882.4
Off-gas (kmol/h)	478.5
Steam to WGS (MMkcal/h)	39.1
Steam to MDEA (MMkcal/h)	37.1
Off-gas export (MMkcal/h)	7.4
Power (MW)	~3.0

increased conversions in both the first and second reactors. It is expected that the heat required for CO_2 sequestration between these two processes would be similar. The coupling of membrane separation modules and the conventional WGSR reactors through this novel architecture results in higher CO conversions and better overall efficiencies. For the same hydrogen product flow of 25,000 Nm^3/h, only

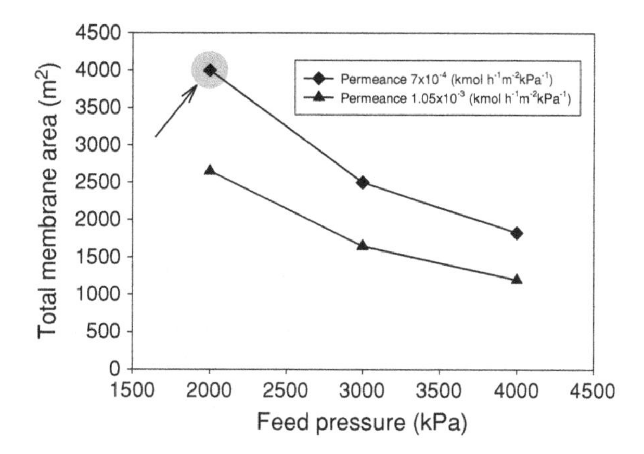

Fig. 7.6 Effect of the inlet feed pressure on the membrane surface area. The *arrow* points at the combination of permeance and pressure used in the case study

about 90% of feed is used compared with the conventional process scheme, resulting in a significant reduction of the variable operating cost.

With the conventional technology, decarbonized fuel is purified by removing the CO_2 from the shifted syngas by low temperature membranes. However, with the OA WGSR membrane reactor, the fuel stream is enriched in H_2 by the membrane reactor and requires only polishing by PSA.

Using the membrane performance listed in Table 7.2, the required membrane surface areas for the two modules have been calculated for the OA scheme, adopting a Hydrogen Recovery Factor (HRF) of 70%. The effect of inlet pressure and H_2 permeance on the membrane surface area is reported in Fig. 7.6 for the first and second modules. The membrane surface area required to achieve the fixed HRF under the conditions dictated by the heat and material balance was calculated using a one-dimensional, steady-state model assuming a steam sweeping ratio of 50%.

7.5 Preliminary Economic Considerations

In order to evaluate the proposed scheme, two options are compared:

1. Conventional 'Pre-combustion decarbonation', where synthesis gas is water–gas shifted to mostly H_2 and CO_2 and then split into two separate streams: one that is CO_2-rich, to be compressed and stored, and one that is pure H_2, to be used as decarbonized fuel. This will be referred to as Scheme A.
2. Membrane-based 'Pre-combustion decarbonation' using the OA scheme as described in Sect. 7.4. This will be referred to as Scheme B.

A detailed comparison of the above two schemes is quite a complex exercise and goes beyond the scope of this study. The purpose of this paragraph is to provide a preliminary economic assessment of the use of membrane reactors to improve the IGCC process in terms of production of a fuel for electric generation and the capture of carbon dioxide.

Table 7.4 Capital investment (CI) comparison between Scheme A and Scheme B

Section	Conventional Scheme A		Membrane-based Scheme B	
	CI[a]	% of CI	CI	% of CI
WGS	3.0	12	6.7[b]	26
CO_2 absorbing and stripping	18.0	72	15.3[c]	60
PSA	4.0	16	3.4[d]	14
Total	25.0 MM€	100	25.5 MM€	100

[a] CI includes all materials, engineering and construction based on KTI data base in MM€ as per a package unit

[b] At least 10% reduction in WGSR reactor volume is expected. For Scheme B, the total membrane surface area of the two membrane modules is estimated at approximately 4,000 m^2 for the worst case scenario

[c] Flow to the absorber is reduced to 60% with unchanged amount of the recovered and stripped CO_2. Since absorber is the largest item, a 15% reduction in CI is expected

[d] Feed flow is reduced by 10% and CO load by 75%. A 5% reduction in the cost is considered

Both schemes share similar concepts; however, the presence of the membrane separator in Scheme B results in a feed requirement reduction of approximately 10%. The membrane-based Scheme B is estimated to have a higher power consumption and higher capital cost for the WGSR than the conventional case but is estimated to have a lower capital cost in the area of CO_2 removal and PSA purification. Owing to the difficulty in the evaluation of the impact of the 10% reduction in feed requirement for a coal unit on the variable costs, our analysis was limited to capital investment. The conventional scheme is considered the base case using heat and material balances developed in the previous section. Table 7.4 compares the capital investment between the conventional case (Scheme A) and the Membrane-based case (Scheme B) for a unit producing 25000 Nm^3/h of pure hydrogen. Since the current study is based mostly on small-scale experiments with membrane reactors and separators, the model presented here is a study in progress. The real system would be very large, within the range of 1500–4000 m^2, as dictated by the combination of membrane permeance and operating pressure of the gasification process. It would also be expensive and probably relatively delicate. It is assumed that such a system can be built and maintained for a membrane cost on the order of 1,000 €/m^2.

As a rough approximation of the investment cost for the membrane-based 'Pre-combustion decarbonation' seems to be aligned with conventional technology if the costs of the membrane and the required membrane surface area remain as predicted. There is a reduction in variable costs associated with the lower feed requirement for the membrane reactor case (Scheme B). This decrease in the cost upstream to the WGSR is estimated to be in the range of 2–3%.

The presently accepted CO_2 penalization foreseen in Europe until 2016 is 26 €/ton of CO_2 captured and removed by a conventional technology. The 10% feed reduction achieved in the OA membrane-based scheme translates into a 28,000-ton-a-year decrease in CO_2 emissions for the plant size under consideration (0.14 kg of CO_2 for each Nm^3 of H_2 produced). It was estimated that the overall average cost of producing hydrogen by WGSR could be reduced by about 4% by applying the OA membrane technology.

For the Post-combustion decarbonation process, CO_2 is removed after combustion at ambient pressure and in a diluted flue gas stream. The amine scrubbing technology in an oxygen environment is still not under control and a compressor is required to increase the flue gas pressure to have a minimum pressure drop on the absorbing system. Since the CO_2 removal section is the most costly in terms of capital investment as shown in Table 7.4, it is reasonable to assume that with an increase of more than 25% in capital investment, the post-combustion decarbonation is going to cost more than the other two processes, without considering the energy penalty associated with compression and solvent regeneration in the O_2 environment. It is, however, obvious that such a comparison is quite rough on a subject which is extremely complex.

The proposed scheme of 'Pre-combustion decarbonation' based on the use of membranes for hydrogen separation is expected to have some advantages over the conventional and commercially ready technology, such as superior efficiency and reduction in overall plant capital cost. Membrane-based systems may represent a real advancement in state-of-the-art H_2 and CO_2 capture in such power plants; however, their real application is closely linked to the realization of a membrane module that is economically built and maintained, as well as to the improvement of membrane permeance and selectivity.

7.6 Conclusion

The OA WGSR membrane reactor scheme examined here is a first step in the evaluation of membrane reactor technology as applied to the IGCC process. The greatest gains, however, are expected with the implementation of the CA configuration where the WGSR and membrane are integrated in a membrane reactor. A substantial plant size reduction could be expected and further process intensification could be achieved by considering other factors such as placement of the catalyst bed within the membrane module, and catalyst structure and functionality. For instance, a foam structured catalyst could enable different flow geometry (radial) along the catalytic bed compared to the conventional one (axial) [70], thus optimizing catalyst effectiveness. Also, the improvement in catalyst functionality may enable operation at lower temperatures without substantially affecting reaction kinetics.

References

1. Burns DT, Piccardi G, Sabbatini L (2008) Some people and places important in the history of analytical chemistry in Italy. Microchim Acta 160:57–87
2. Mond L, Langer C (1888) Improvements in obtaining hydrogen. British Patent 12608

3. Navarro RM, Peña MA, Fierro JLG (2007) Hydrogen production reactions from carbon feedstocks: fossil fuels and biomass. Chem Rev 107:3952–3991
4. Lei Y, Cant NW, Trimm DL (2005) Activity patterns for the water gas shift reaction over supported precious metal catalysts. Catal Lett 103:133–136
5. Lei Y, Cant NW, Trimm DL (2005) Kinetics of the water–gas shift reaction over a rhodium-promoted iron–chromium oxide catalyst. Chem Eng J 114:81–85
6. Lei Y, Cant NW, Trimm DL (2006) The origin of rhodium promotion of Fe_3O_4–Cr_2O_3 catalysts for the high-temperature water–gas shift reaction. J Catal 239:227–236
7. Rhodes C, Williams BP, King F, Hutchings GJ (2002) Promotion of Fe_3O_4/Cr_2O_3 high temperature water gas shift catalyst. Catal Commun 3:381–384
8. Natesakhawat S, Wang X, Zhang L, Ozkan US (2006) Development of chromium-free iron-based catalysts for high-temperature water-gas shift reaction. J Mol Catal A Chem 260:82–94
9. Martos C, Dufour J, Ruiz A (2009) Synthesis of Fe_3O_4-based catalysts for the high-temperature water gas shift reaction. Int J Hydrogen Energy 34:4475–4481
10. Maroño M, Ruiz E, Sánchez JM, Martos C, Dufour J, Ruiz A (2009) Performance of Fe–Cr based WGS catalysts prepared by co-precipitation and oxi-precipitation methods. Int J Hydrogen Energy 34:8921–8928
11. Nishida K, Atake I, Li D, Shishido T, Oumi Y, Sano T, Takeira K (2008) Effects of noble metal-doping on $Cu/ZnO/Al_2O_3$ catalysts for water–gas shift reaction. Catalyst preparation by adopting "memory effect" of hydrotalcite. Appl Catal A 337:48–57
12. Guo P, Chen L, Yang Q, Qiao M, Li H, Li H, Xu H, Fan K (2009) $Cu/ZnO/Al_2O_3$ water–gas shift catalysts for practical fuel cell applications: the performance in shut-down/start-up operation. Int J Hydrogen Energy 34:2361–2368
13. Tang X-J, Fei J-H, Hou Z-Y, Lou H, Zheng X-M (2008) Copper-zinc oxide and manganese promoted copper-zinc oxide as highly active catalysts for water–gas shift reaction. React Kinet Catal Lett 94:3–9
14. Li Y, Fu Q, Flytzani-Stephanopoulos M (2000) Low-temperature water-gas shift reaction over Cu- and Ni-loaded cerium oxide catalysts. Appl Catal B 27:179–191
15. Djinović P, Batista J, Pintar A (2008) Calcination temperature and CuO loading dependence on CuO–CeO_2 catalyst activity for water-gas shift reaction. Appl Catal A 347:23–33
16. Djinović P, Levec J, Pintar A (2008) Effect of structural and acidity/basicity changes of CuO–CeO_2 catalysts on their activity for water–gas shift reaction. Catal Today 138:222–227
17. Djinović P, Batista J, Levec J, Pintar A (2009) Comparison of water–gas shift reaction activity and long-term stability of nanostructured CuO–CeO_2 catalysts prepared by hard template and co-precipitation methods. Appl Catal A 364:156–165
18. Djinović P, Batista J, Pintar A (2009) WGS reaction over nanostructured CuO–CeO_2 catalysts prepared by hard template method: characterization, activity and deactivation. Catal Today 147S:S191–197
19. Wheeler C, Jhalani A, Klein EJ, Tummala S, Schmidt LD (2004) The water–gas-shift reaction at short contact times. J Catal 223:191
20. Radhakrishnan R, Willigan RR, Dardas Z, Vanderspurt TH (2006) Water gas shift activity and kinetics of Pt/Re catalysts supported on ceria-zirconia oxides. Appl Catal B 66:23–28
21. Jacobs G, Ricote S, Davis BH (2006) Low temperature water-gas shift: type and loading of metal impacts decomposition and hydrogen exchange rates of pseudo-stabilized formate over metal/ceria catalysts. Appl Catal A 302:14–21
22. Panagiotopoulou P, Papavasiliou J, Avgouropoulos G, Ioannides T, Kondarides DI (2007) Water–gas shift activity of doped Pt/CeO_2 catalysts. Chem Eng J 134:16–22
23. Kim YT, Park ED, Lee HC, Lee D, Lee KH (2009) Water–gas shift reaction over supported Pt-CeO_x catalysts. Appl Catal B 90:45–54
24. Lim S, Bae J, Kim K (2009) Study of activity and effectiveness factor of noble metal catalysts for water–gas shift reaction. Int J Hydrogen Energy 34:870–876
25. Ciambelli P, Palma V, Palo E, Iaquaniello G (2009) Experimental and economical approach to the integration of a kW-scale CH_4-ATR reactor with a WGS stage. Chem Eng Trans 18:499–504

26. Duarte de Farias AM, Nguyen-Thanh D, Fraga MA (2010) Discussing the use of modified ceria as support for Pt catalysts on water–gas shift reaction. Appl Catal B 93:250–258
27. González ID, Navarro RM, Wen W, Marinkovic N, Rodriguéz JA, Rosa F, Fierro JLG (2010) A comparative study of the water gas shift reaction over platinum catalysts supported on CeO_2, TiO_2 and Ce-modified TiO_2. Catal Today 149:372–379
28. Fu Q, Weber A, Flytzani-Stephanopoulos M (2001) Nanostructured Au–CeO_2 catalysts for low-temperature water–gas shift. Catal Lett 77:87–95
29. Fu Q, Kudriavtseva S, Saltsburg H, Flytzani-Stephanopoluos M (2003) Gold-ceria catalysts for low-temperature water–gas shift reaction. Chem Eng J 93:41–53
30. Burch R (2006) Gold catalysts for pure hydrogen production in the water-gas shift reaction: activity, structure and reaction mechanism. Phys Chem Chem Phys 8:5483–5500
31. Andreeva D, Ivanov I, Ilieva L, Sobczak JW, Avdeev G, Petrov K (2007) Gold based catalysts on ceria and ceria-alumina for WGS reaction (WGS Gold catalysts). Top Catal 44:173–182
32. Sandoval A, Gómez-Cortés A, Zanella R, Díaz G, Saniger JM (2007) Gold nanoparticles: support effects for the WGS reaction. J Mol Catal A 278:200–208
33. Fonseca AA, Fisher JM, Ozkaya D, Shannon MD, Thompsett D (2007) Ceria-zirconia supported Au as highly active low temperature water–gas shift catalysts. Top Catal 44:223–235
34. Bond G (2009) Mechanisms of the gold-catalysed water-gas shift. Gold Bull 42:337–342
35. Personal Communications with Elcogas-MRP-Cosero, October (2006)
36. Bracht M, Alderliesten PT, Kloster R, Pruschek R, Haupt G, Xue E, Ross JRH, Koukou MK, Papayannakos N (1996) Water gas shift membrane reactor for CO_2 control in IGCC systems: techno-economic feasibility study. Energy Convers 38:S159–164
37. Chiesa P, Kreutz TG, Lozza GG (2007) CO_2 sequestration from IGCC power plants by means of metallic membranes. J Eng Gas Turbines Power 129:123–134
38. De Lorenzo L, Kreutz TG, Chiesa P, Williams RH (2008) Carbon-free hydrogen and electricity from coal: options for syngas cooling in systems using a hydrogen separation membrane reactor. J Eng Gas Turbines Power 130:031401-1
39. Diniz da Costa JC, Reed GP, Thambimuthu K (2009) High temperature gas separation membranes in coal gasification. Energy Procedia 1:295–302
40. Manzolini G, Viganò F (2009) Co-production of hydrogen and electricity from autothermal reforming of natural gas by means of Pd-Ag membranes. Energy Procedia 1:319–326
41. Rezvani S, Huang Y, McIlveen-Wright D, Hewitt N, Mondol JD (2009) Comparative assessment of coal fired IGCC systems with CO_2 capture using physical absorption, membrane reactors and chemical looping. Fuel 88:2463–2472
42. Jansen D, Dijkstra JW, van den Brink RW, Peters TA, Stange M, Bredesen R, Goldbach A, Xu HY, Gottschalk A, Doukelis A (2009) Hydrogen membrane reactors for CO_2 capture. Energy Procedia 1:253–260
43. Website: http://www.sud-chemie.com
44. Doong SJ, Lau F, Roberts M, Ong E (2005) GTI's solid fuel gasification to hydrogen program. In: The 3rd natural gas technology conference, Orlando, FL
45. Bredesen R, Jordal K, Bolland O (2004) High-temperature membranes in power generation with CO_2 capture. Chem Eng Process 43:1129–1158
46. Basile A, Drioli E, Santella F, Violante V, Capannelli G, Vitulli G (1995) A study on catalytic membrane reactors for water gas shift reaction. Gas Sep Purif 10:53–61
47. Basile A, Criscuoli A, Santella F, Drioli E (1996) Membrane reactor for water gas shift reaction. Gas Sep Purif 10:243–254
48. Criscuoli A, Basile A, Drioli E (2000) An analysis of the performance of membrane reactors for the water–gas shift reaction using gas feed mixtures. Catal Today 56:53–64
49. Basile A, Chiappetta G, Tosti S, Violante V (2001) Experimental and simulation of both Pd and Pd/Ag for a water gas shift membrane reactor. Sep Purif Technol 25:549–571
50. Iyoha O, Enick R, Killmeyer R, Howard B, Morreale B, Ciocco M (2007) Wall-catalyzed water-gas shift reaction in multi-tubular Pd and 80 wt% Pd–20 wt% Cu membrane reactors at 1173 K. J Membr Sci 298:14–23

51. Iyoha O, Enick R, Killmeyer R, Howard B, Ciocco M, Morreale B (2007) H_2 production from simulated coal syngas containing H_2S in multi-tubular Pd and 80 wt% Pd–20 wt% Cu membrane reactors at 1173 K. J Membr Sci 306:103–115
52. Barbieri G, Brunetti A, Tricoli G, Drioli E (2008) An innovative configuration of a Pd-based membrane reactor for the production of pure hydrogen. Experimental analysis of water gas shift. J Power Sources 182:160–167
53. Bi Y, Xu H, Li W, Goldbach A (2009) Water–gas shift reaction in a Pd membrane reactor over $Pt/Ce_{0.6}Zr_{0.4}O_2$ catalyst. Int J Hydrogen Energy 34:2965–2971
54. Brunetti A, Barbieri G, Drioli E (2009) Upgrading of a syngas mixture for pure hydrogen production in a Pd–Ag membrane reactor. Chem Eng Sci 64:3448–3454
55. Brunetti A, Barbieri G, Drioli E (2009) Pd-based membrane reactor for syngas upgrading. Energy Fuels 23:5073–5076
56. Brunetti A, Barbieri G, Drioli E, Granato T, Lee K-H (2007) A porous stainless steel supported silica membrane for WGS reaction in a catalytic membrane reactor. Chem Eng Sci 62:5621–5626
57. Giessler S, Jordan K, Diniz da Costa JC, Lu GQ(M) (2003) Performance of hydrophobic and hydrophilic silica membrane reactors for the water gas shift reaction. Sep Purif Technol 33:255–264
58. Battersby S, Duke MC, Liu S, Rudolph V, Diniz da Costa JC (2008) Metal doped silica membrane reactor: operational effects of reaction and permeation for the water gas shift reaction. J Membr Sci 316:46–52
59. Battersby S, Smart S, Ladewig B, Liu S, Duke MC, Rudolph V, Diniz da Costa JC (2009) Hydrothermal stability of cobalt silica membranes in a water gas shift membrane reactor. Sep Purif Technol 66:299–305
60. Kikuchi E, Uemiya S, Sato N, Inoue H, Ando H, Matsuda T (1989) Membrane reactor using microporous glass supported thin film of palladium. Application to the water gas shift reaction. Chem Lett 18:489–492
61. Uemiya S, Sato N, Ando H, Kikuchi E (1991) The water gas shift reaction assisted by a palladium membrane reactor. Ind Eng Chem Res 30:585–589
62. Criscuoli A, Basile A, Drioli E, Loiacono O (2001) An economic feasibility study for water gas shift membrane reactor. J Membr Sci 181:21–27
63. Galuszka J, Giddings T (accepted for publication) Silica membranes-preparation by chemical vapour deposition and characteristics. In: Basile A (ed) Membranes for membrane reactors: preparation, optimization and selection, chap 12. Wiley
64. Brunetti A, Barbieri G, Drioli E, Lee K-H, Sea B, Lee D-W (2007) WGS reaction in a membrane reactor using a porous stainless steel supported silica membrane. Chem Eng Process 46:119–126
65. Galuszka J, Giddings T, Iaquaniello G. Integration of membrane reactor and IGCC technologies: experimental study and reactor modeling. Chem Eng Process (to be published)
66. Li J, Yoon H, Oh TK, Wachsman ED (2009) High temperature $SrCe_{0.9}Eu_{0.1}O_{3-\delta}$ proton conducting membrane reactor for H_2 production using the water–gas shift reaction. Appl Catal B 92:234–239
67. Huang J, El-Azzami L, Ho WSW (2005) Modeling of CO_2-selective water gas shift membrane reactor for fuel cell. J Membr Sci 261:67–75
68. Zou J, Huang J, Ho WSW (2007) CO_2-selective water gas shift membrane reactor for fuel cell hydrogen processing. Ind Eng Chem Res 46:2272–2279
69. Huang J, Ho WSW (2008) Effect of system parameters on the performance of CO_2-selective WGS membrane reactor for fuel cells. J Chin Inst Chem Eng 39:129–136
70. Palma V, Palo E, Ciambelli P (2009) Structured catalytic substrates with radial configurations for the intensification of the WGS stage in H_2 production. Catal Today 147S:S107–112

Chapter 8
Membrane-Assisted Catalytic Cracking of Hydrogen Sulphide (H_2S)

Jan Galuszka, Gaetano Iaquaniello, Paolo Ciambelli, Vincenzo Palma and Elvirosa Brancaccio

8.1 Introduction

Hydrogen sulphide (H_2S) occurs naturally in many gas wells; its concentration in natural gas varies from traces to 90% by volume. E.g. Canada's Caroline and Bearberry gas fields in West-Central Alberta contain between 70 and 90% of H_2S. In the industry, H_2 is a by-product of sour natural gas sweetening, hydrodesulphurization of light hydrocarbons, and upgrading of heavy oils, bitumen and coals. Since this H_2S has a limited industrial application, it is viewed as a pollutant requiring treatment and removal. Hydrogen sulphide has a high heating value, but its use as a fuel is ruled out because sulphur dioxide is a product of H_2S combustion, which is environmentally unacceptable. At present, H_2S is separated from hydrocarbon gases by amine adsorption and regeneration, producing acid gas containing 10–90% by volume of H_2S. When concentrations of H_2S exceed 40%, this gas is treated by the Claus process in which H_2S is oxidized to water and

J. Galuszka (✉)
Natural Resources Canada, CanmetENERGY, 1 Haanel Drive,
Ottawa, ON K1A 1M1 Canada
e-mail: galuszka@NRCan.gc.ca

G. Iaquaniello
Tecnimont KT S.p.A, Viale Castello della Magliana 75,
00148 Rome, Italy
e-mail: Iaquaniello.G@tecnimontkt.it

P. Ciambelli and V. Palma
Department of Chemical and Food Engineering, University of Salerno,
84084 Fisciano, SA, Italy
e-mail: pciambelli@unisa.it

E. Brancaccio
Processi Innovativi S.r.l., Corso Federico II 36, 67100 L'Aquila, Italy
e-mail: ebrancaccio@unisa.it

M. De Falco et al. (eds.), *Membrane Reactors for Hydrogen Production Processes*,
DOI: 10.1007/978-0-85729-151-6_8,

sulphur. This process is uneconomical when the price of sulphur (the primary product) is depressed; nevertheless, it is done to dispose of H_2S in an environmentally acceptable manner.

However, H_2S has a much higher economic value if the hydrogen, as well as the sulphur, could be recovered. Refineries use hydrogen as a basic reagent in hydrocracking and in hydrotreating, to produce fuels with low sulphur and low aromatics content. Hydrogen is also used in the chemical industry, primarily in the synthesis of ammonia and methanol. Most of the hydrogen currently consumed is manufactured by steam methane reforming. This process requires large quantities of natural gas as both the feed gas and combustion fuel, resulting in large amounts of carbon dioxide being produced and discharged into the atmosphere, thus contributing significantly to the 'greenhouse' effect [1].

Refining oil sands also produces a significant amount of H_2S; here again, the Claus process is used, yielding sulphur and water. It is preferable to convert H_2S to sulphur and hydrogen within the refinery and return the recovered hydrogen to the heavy oil hydrogenation step. This would significantly improve hydrogen inventory and reduce associated CO_2 emissions. For instance, the bitumen mined from the oil sands deposits in the Athabasca area of Alberta, Canada, contains 4.4 wt% of sulphur. At present, conversion of this sulphur to H_2S requires about 0.2 Mt/year of hydrogen. It is projected that by 2015, the amount of hydrogen wasted to H_2S will increase to 0.6 Mt/year, as the processing of bitumen will reach 3.261 Mbbl/day [2]. Replacement of that hydrogen by steam reforming of natural gas is currently responsible for about 0.8 Mt/year of CO_2 emissions. By 2015, emissions will increase to 3 Mt/year of CO_2.

Many processes were studied and proposed for the decomposition of H_2S including thermal [3–7] and thermochemical [8, 10–15, 57] decomposition; photochemistry [16–24]; plasma methods [25–27, 67]; electrolysis [9, 28–30]; and electrochemical cycles [31–33]. These approaches are at various stages of development covering numerous disciplines. However, none has yet been implemented due to either cost or technical feasibility. This has prompted substantial research in H_2S decomposition with equilibrium shift by several methods such as preferential removal of reaction products by membranes [12, 34–53] and thermal diffusion columns [50–57]. The common drawback of all these proposed solutions is the requirement for an external energy supply. Energy production, in most cases, would cause CO_2 emissions defying the concept of H_2S being the source of clean hydrogen. Also, any future technology developed to convert H_2S to sulphur and hydrogen must compete, both technically and economically, with current hydrogen production processes and, simultaneously, compare favourably with the Claus process for sulphur production.

In this chapter, the most popular approaches for hydrogen production from H_2S are reviewed before a novel open reactor architecture (OA) is presented, where the coupling of reaction and hydrogen separation is achieved in the series of consecutive conventional catalytic reactors (CR), each followed by a membrane separator (MS). Experimental study on the development of a suitable H_2S decomposition catalyst is also presented, and the theoretical calculations for one

CR/MS/CR unit predicting an overall one-pass H_2S conversion close to 40% at ambient pressure are discussed. Finally, a preliminary costing of the proposed scheme that allows hydrogen production without CO_2 emissions is summarized.

8.2 Thermodynamic and Reaction Kinetics Considerations

Hydrogen sulphide decomposition is a reversible reaction controlled by thermodynamic equilibrium. The reaction is highly endothermic; as shown in Fig. 8.1, the predicted conversion of thermally decomposed H_2S based on thermodynamic equilibrium is only about 20% at 1000°C and 38% at 1200°C. Temperatures exceeding 1375°C are needed to drive the H_2S decomposition reaction to conversions above 50%.

Since the number of moles increases during H_2S decomposition, the pressure has a negative effect on conversion as shown in Fig. 8.2 and also reported by Faraji et al. [58].

Also, the conversion is negatively influenced by the presence of one or both products of H_2S decomposition in the reaction mixture, especially H_2, as shown in Fig. 8.3.

At higher concentrations the hydrogen presence suppresses the H_2S conversion more significantly than sulphur. Consequently, removal of hydrogen from the reaction mixture should be more beneficial for the process of H_2S dissociation.

The kinetics of H_2S dissociation reaction were studied extensively [3, 4, 7, 59–64]. Although the decomposition to S_2 and H_2 has often been used as the representative reaction, other sulphur allotropes, sulphanes and HS are also formed and need to be considered, especially at lower temperature, as recognized by Raymont [3]. Detailed high-temperature kinetics of H_2S decomposition pertinent to forward and reverse part of the reaction were reported by Kaloidas et al. [4] and Dowling et al. [59], respectively. Although the estimated

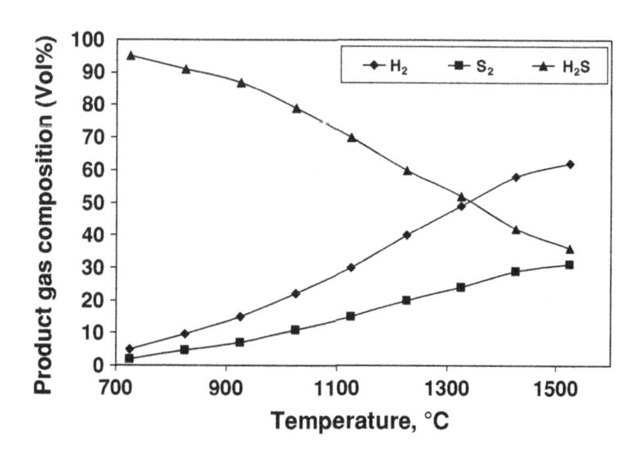

Fig. 8.1 Product gas composition calculated for the pure H_2S decomposition reaction as a function of temperature

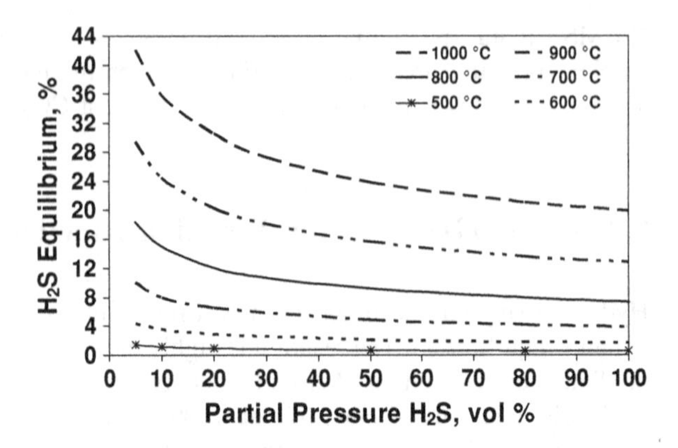

Fig. 8.2 H_2S partial pressure effect on equilibrium, calculated using Gaseq (Chemical equilibria in perfect gases)

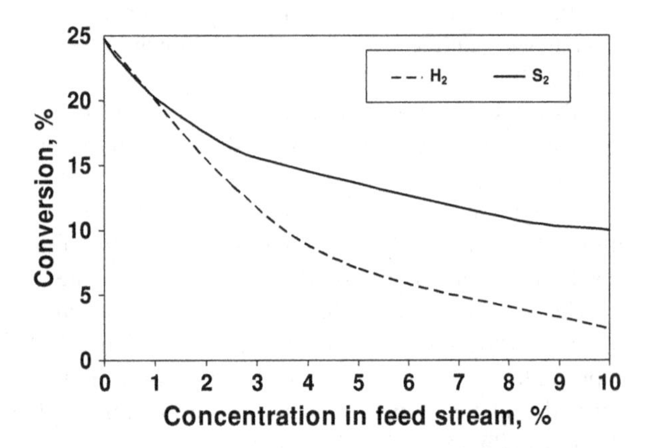

Fig. 8.3 Dependence of H_2S conversion on H_2 or S_2 presence in the feed mixture at 900°C, calculated using Gaseq (Chemical equilibria in perfect gases)

values of the apparent activation energy were similar, there was a significant disagreement between the value of the Arrenhius factors and the order of the rate equation found by these two investigations, leading to a difference of roughly 25% between predicted conversions of dissociation and re-association reactions. More recently, Hawboldt et al. [61] determined a new rate expression for H_2S dissociation/re-association reaction. The proposed rate expression is

$$r = A_f e^{-E_f/RT} P_{H_2S} P_{S_2}^{0.5} - A_r e^{-E_r/RT} P_{H_2} P_{S_2} \quad (8.1)$$

where A_f and E_f are 5260 (±260) mol/(cm^3 s atm$^{1.5}$) and 45.0 (±0.3) kcal/mol, respectively, and A_r and E_r are 14 (±1) mol/(cm^3 s atm^2) and 23.6 (±0.2) kcal/mol.

It was determined that at temperatures below 1000°C and residence durations below 500 ms, the rate of H_2S dissociation was insignificant. Furthermore, overall conversions of H_2S are low, even at long residence durations, at temperatures below 950°C.

8.3 H_2S Decomposition Strategies

8.3.1 Thermal

It is widely recognized that the most direct process for converting H_2S to hydrogen and sulphur is catalytic or non-catalytic thermal decomposition [3–7]. However, due to thermodynamic restrictions and the endothermic character of H_2S dissociation reaction, this approach is considered impractical as temperatures exceeding 1000°C are needed to overcome these hurdles and achieve an industrially interesting conversion. In addition, an unreacted H_2S needs to be recycled to maximize H_2 yield, and the product gases must be separated by rapid quenching to block the recombination of H_2 and S_2. Consequently, the cost of such a process is prohibitive due to energy and exotic metallurgy requirements.

8.3.2 Thermochemical

The high energy requirement for H_2S dissociation could be circumvented by a series of less energy intensive steps involving sulphiding of a metal, or a lower sulphide to liberate the hydrogen and then calcining the higher sulphide to decompose into sulphur and the original metal or sulphide, as described by Chivers et al. [8]. The use of alkali metal sulphides and polysulphides were studied by Chivers and Lau [13]. Vanadium sulphide cycles were studied by Al-Shamma and Naman [10] and Chivers and lau [57]. Although the achieved hydrogen yields were higher when compared to direct thermal decomposition, it should be noted that polysulphides are very corrosive and difficult to handle above 650°C. Also, it was observed by Luinstra [11] that the diffusion mass transfer limits the efficiency of the metal sulphide-based processes.

Open loop thermochemical cycles that employ carbon oxides or hydrocarbons were proposed by Raymont [12], but gas separation was problematic.

Marathon Oil developed a thermochemical process based on the oxidizing ability of *t*-butylanthraquinone [14] and further studied by Mark et al. [15] who employed computational chemistry using semi-empirical methods to determine the detailed chemical steps of the process occurring through S_8 formation.

At this stage, none of the thermochemical cycles described in the literature has reached commercial development.

8.3.3 Electrochemical

Gregory et al. [9] concluded that because of low specific conductance and a low dielectric constant of liquid H_2S, production of hydrogen by direct electrolysis seems impractical. Electrolysis in water solution was studied by Bolmer [28, 29] and Johnson [30]. Since it was determined that the number of ionic species in H_2S water solution was too small to obtain satisfactory rates of H_2S dissociation, a supporting acidic or basic electrolytes were used to raise the conductance of the solution. However, elemental sulphur produced by the electrolysis is a good insulator, causing a substantial increase in the power demand and lowering the electrolysis efficiency with time, which makes this approach impractical.

Indirect electrolysis of H_2S using acidic or basic iodine to oxidize H_2S was studied by Kalina et al. [31] However, the acidic process suffered from loss of iodine and formation of impure, sticky plastic sulphur. The basic process gave low sulphur yields but had excessive oxidant consumption and high electrical energy requirements.

The direct or indirect electrolysis of H_2S produces high purity hydrogen, but the high cost of electricity makes these methods uneconomical at the present time for hydrogen production on a commercial scale.

8.3.4 Solar

The use of solar energy for experimental studies on H_2S uncatalysed thermolysis indicated that a high degree of chemical conversion was attainable and that the reverse reaction during quench was negligible [6, 65, 66]. The recommended process temperature would be between 1300 and 1500°C. However, a preliminary economic assessment indicated that the capital cost of a plant employing effusion to separate H_2 and H_2S would be up to ten times higher than a similar capacity Claus plant.

8.3.5 Photochemical

Most attempts to dissociate H_2S photolytically involve irradiating an alkaline solution of the gas in which semiconductor particles are suspended. Cadmium sulphide, ruthenium dioxide/cadmium sulphide, titanium dioxide, chromium, platinum, and vanadium sulphides were used [16–18] as the semiconductor which participates by absorbing the incident quanta of light.

A recent review [19, 22] on stratified photocatalyst CdS (nanoparticles arranged into capsule form), catalyzing the decomposition of H_2S into H_2 and S claimed that H_2S can be directly split into hydrogen and sulphur on photocatalysts composed of CdS-based semiconductors loaded with noble metals and noble metal sulphides, using nonaqueous ethanolamine solvent as the reaction media. The quantum efficiency in hydrogen production can be as high as 30% under visible light irradiation. None of these materials improved hydrogen production significantly, and the efficiency of processes using light to dissociate H_2S to hydrogen and sulphur was approximately 3%.

8.3.6 Plasma

The use of plasma reactors to dissociate H_2S was reported by Argyle et al. [25]. It was claimed that the energy efficiency of the processes was low, presumably because many successive dissociation–recombination processes served only to recreate the reactant H_2S and produced heat before H_2 is finally formed as a product. The calculated energy requirements of 0.5–200 eV/molecule of H_2S converted was many times higher than the theoretical minimum of $\sim$0.2 eV/ molecule H_2S (21 kJ/mol H_2S) based on the enthalpy of formation of H_2S at 298 K.

Breakdown voltages of H_2S in four balance gases (Ar, He, N_2 and H_2) were studied by Zhaoa et al. [67] who concluded that no thermal plasmas are effective for dissociating H_2S into hydrogen and sulphur, but further increases in energy efficiency are necessary. It was emphasized by Thomas [26] that an understanding of the reaction parameter space, which includes pulse frequency, discharge capacitance and voltage, reactor residence time and electrode material, is vital to optimize reactor performance and energy efficiency.

Until now, plasmas have not been used for any large scale chemical processing. The biggest obstacle possibly is the high consumption of electrical energy. However, the plasma process is environmentally friendly and operationally simple with minimum waste generation.

8.3.7 Microwaves

Application of microwave radiation to heterogeneous catalytic systems was reviewed by Roussy and Pearce [68]. Comparison of the effect of microwave and conventional heating on catalytic decomposition of H_2S into hydrogen and sulphur, carried out on impregnated and mechanically mixed MoS_2/γ-alumina (both 30% by weight MoS_2) was investigated by Zhang and Hayward [69, 70]. It was concluded that the calculated equilibrium conversion based on thermodynamic data of Kaloidas and Papayannakas [71] for the conventional thermal reaction correlated with the equilibrium data, whereas the reaction carried out in the microwave cavity achieved conversion higher than that expected by thermodynamics.

8.3.8 Membrane-Assisted Conversion

Thermal catalytic decomposition of H_2S to hydrogen and sulphur is a good candidate for application of a membrane reactor. The expected significant process simplification and intensification would capitalize on a new industrial paradigm offered by membrane reactors that allows combining reaction and separation in one step. A common feature of all catalytic membrane reactors having a hydrogen permselective membrane wall is the ability of such a reactor to circumvent thermodynamic limitations of an equilibrium-controlled process, by separating H_2 as it is produced. Therefore, it is expected that membranes with high selectivity and permeability towards hydrogen—if successfully integrated with advanced catalysts—would make H_2S dissociation commercially feasible, by enabling hydrogen production at a much lower temperature with greater intensity.

Both ceramic and composite metal hydrogen permselective membranes were considered for the separation of hydrogen from H_2S. However, the practical application of these membranes has some restrictions. The noble metal membranes are affected by chemical attack by H_2S and their use entails considerable cost. Separation by mesoporous membranes governed by Knudsen diffusion (pore size 2–50 nm) provides insufficient selectivity towards hydrogen. Microporous silica hydrogen permselective membranes (H-membranes) have also been extensively studied [72] as a potentially more practical alternative to Pd membranes. Chemical vapour deposition (CVD) and the sol–gel method are the most commonly used techniques for fabricating silica membranes. These state-of-the-art silica membranes have good hydrogen flux and separation and respectable thermal stability. However, hydrothermal stability of silica H-membrane is a key factor in determining its suitability for a commercial application for membrane-assisted processes [72].

The application of a membrane reactor with a hydrogen permselective membrane to intensify H_2S decomposition to hydrogen and sulphur at a lower temperature was first considered in the eighties [50, 73, 74]. However, the expected positive effects on H_2S conversion observed with the early Vycor-based membranes were insignificant. Also, a zirconia–silica membrane developed more recently and applied to decomposition of H_2S by Ohashi et al. [75] had only Knudsen selectivity and produced similar results.

A catalytic membrane reactor having a tubular ceramic membrane for H_2S decomposition was patented by Vizoso [76]. It was claimed that applying a membrane reactor between 400 and 700°C, with the molybdenum sulphide catalyst deposited directly on the surface of the ceramic membrane, resulted in a 20% increase in the conversion of a 4% H_2S stream, though few details were provided.

The most recent application of a silica membrane—prepared by counter-diffusion CVD of TMOS and oxygen—to H_2S decomposition was reported by Akamatsu et al. [77]. With this membrane and a commercial desulphurization catalyst at 600°C, about 70% H_2S diluted in 99% nitrogen was converted, in a relatively short residence time of 7 s. Nonetheless, this claim is not supported by a material balance, and the amount of hydrogen passing through the membrane in a very diluted feed stream is not reported.

A multilayer metallic membrane reactor for H_2S decomposition was patented by Edlund and Friesen [78] and, later, further described by Edlund and Pledger [79, 80]. The membrane consisted of three distinct layers: a base-metal/mechanical support layer, an intermediate layer and a coating-metal layer. The coating metal was a H_2-permeable metal resistant to chemical corrosion by H_2S (e.g. Pd alloyed with about 40% Cu). It was reported that at 700°C, the membrane-assisted thermolysis of 1.5% of H_2S in nitrogen under a total pressure of about 8 atm was practically driven to completion, whereas the equilibrium conversion under similar conditions without hydrogen removal was only 13%. However, the required contact time to reach this conversion was about 12 min, which is impractical.

The practicality of Pd membrane application to H_2S decomposition was the subject of several studies. Morreale et al. [37] evaluated the effect of H_2S on the hydrogen permeance of Pd–Cu alloys at temperatures above 600°C. It was claimed that the face-centred cubic stability region of the Pd–Cu alloy is least affected by the presence of 1000 ppm H_2S. The greatest effect of H_2S on hydrogen permeance occurred at the lowest temperatures of about 300°C, which corresponded to the body-centred cubic crystalline phase of the alloy. This agrees with other studies [38–40], suggesting that at the high temperature some Pd–Cu compositions may tolerate impurity levels of 1000 ppm H_2S.

Recently, Pd–Nb alloys were investigated by Aboud et al. [41]. Niobium shows resistance to sulphur poisoning and has the highest hydrogen permeability that could be an order of magnitude higher than pure Pd. Unfortunately, it was determined that Nb is notorious for undergoing hydrogen embrittlement and its presence can be a major structural issue.

Clearly, further development and newer strategies are needed before a successful and practical application of a membrane reactor to hydrogen production through catalytic H_2S decomposition can be claimed.

8.4 H_2S Decomposition Catalyst Development

Owing to the inherent slowness of H_2S thermal decomposition (about 15 s is needed to reach steady-state conditions at 1100°C [3]), substantial effort was made to develop a catalyst that would promote decomposition reaction at lower temperatures. Since H_2S, even at very low concentrations in the ppm range, exhibits a strong poisoning effect on many metal-based catalysts [81], hydrodesulphuration catalysts known for their resistance to H_2S poisoning were considered and investigated for H_2S decomposition. Among these, molybdenum disulphide and tungsten disulphide were found to be very effective for the decomposition of H_2S. The following activity order determined experimentally in the temperature range of 180–420°C was reported [12, 82, 83]:

$$CoS_2 > NiS = WS_2 > MoS_2 > FeS_2 > Ag_2S > CuS > CdS > MnS > ZnS.$$

Chivers et al. [8] compared the catalytic activity of some metal sulphide powders, and for the series Cr_2S_3, MoS_2 and WS_2, it was found that MoS_2 is the most active catalyst at $T \geq 600°C$, but Cr_2S_3 and WS_2 demonstrated greater yield of hydrogen at $T \leq 600°C$ than MoS_2.

The catalytic activity tests for H_2S decomposition carried out at the Chemical and Food Engineering Department of the University of Salerno (UNISA), showed interesting results at a relatively low temperature of 900°C and in a practical range of gas hourly space velocity (GHSV) between 5000 and 100000 h^{-1}. Quartz chips as inert and four different catalyst formulations were tested that included Al_2O_3 fresh and calcined at 800°C, MoS_2 powder, and alumina supported Pt and MoS_2 between 700 and 1000°C and in the gas hourly space velocity (GHSV) range of 20000–900000 h^{-1}. The total feed flow rate was between 0.25 and 1.2 l/min (STP), and H_2S was diluted in N_2 giving compositions between 5 and 100% at atmospheric pressure.

Figure 8.4 shows H_2S conversions as a function of temperature for the feed stream containing 10% H_2S in N_2. It is clear that at the GHSV of 50000 h^{-1} in the temperature range of 600–1000°C all the studied catalysts were able to accelerate the H_2S decomposition. In particular, both the MoS_2 supported on alumina and the pure calcined alumina gave conversions close to equilibrium, although, in the temperature range of 700–900°C, the MoS_2/Al_2O_3 performed better.

Figure 8.5 shows the gas hourly space velocity (GHSV) influence on H_2S conversions studied at 800 and 900°C for the feed stream containing 10% H_2S in N_2 tested for the same selection of catalysts. Since the H_2S conversions with quartz chips were below 2%, they were not included in Fig. 8.5.

The 10% MoS_2/Al_2O_3 catalyst showed the best performance at 900°C, as H_2S conversions were in a narrow range between 21 and 25% for a wide range of GSHV between 5000 and 700000 h^{-1}. The lowest contact time was about 5 ms corresponding to GHSV of 700000 h^{-1}. All the other catalysts were able to bring the system to near equilibrium but in a much narrower range of GHSV.

At 800°C, again, the 10% MoS_2/Al_2O_3 catalyst gave the best results but equilibrium conversion was not reached. At GHSV of 18000 h^{-1} (contact time of 0.2 s) the conversion was 14%. At the longest contact time below GHSV of 5000 h^{-1}, the observed conversion decrease is caused most likely by the recombination reaction between hydrogen and sulphur, as these products of H_2S decomposition were not removed fast enough from the reaction zone. It was verified that during 20 h on stream, the deactivation of the 10% MoS_2/Al_2O_3 catalyst was not detectable and, at 900°C and contact time of 5 ms, it gave a desirable 25% of H_2S conversion.

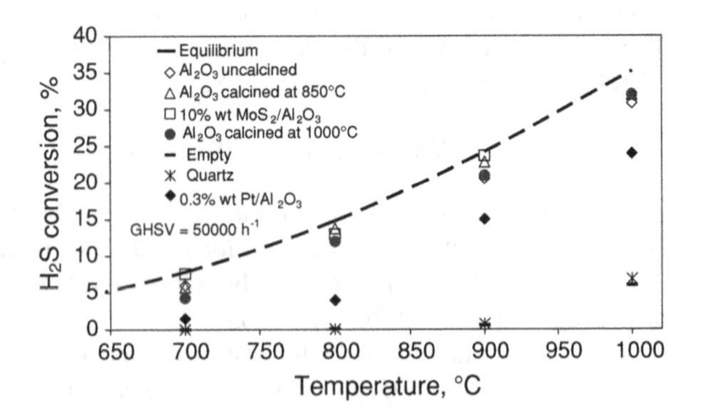

Fig. 8.4 Influence of temperature on H_2S decomposition tested with various catalysts

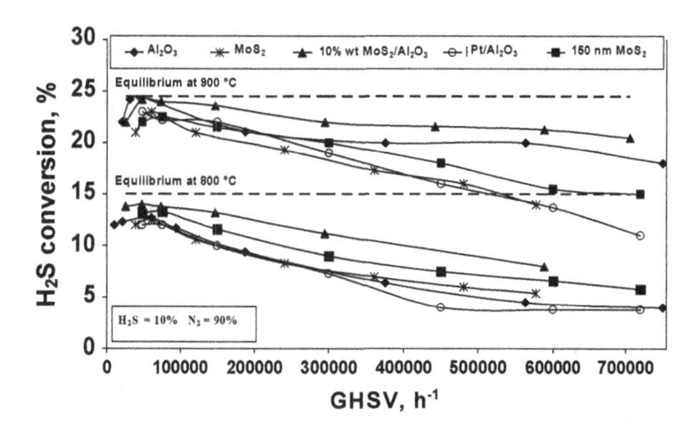

Fig. 8.5 Influence of contact time on catalytic decomposition of H_2S tested for various catalysts

8.5 Novel Process Configuration

The proposed process configuration for a membrane-assisted decomposition of H_2S to hydrogen and sulphur is shown schematically in Fig. 8.6. The single section contains two conventional CRs and one MS placed between them, forming the so-called open membrane reactor architecture (OA). Therefore, the coupling of reaction and hydrogen separation would be achieved in a series of the consecutive CR each followed by the MS unit (CRMS). The number of the CRMS units would be determined by the required feed conversion. Such membrane-assisted reaction architecture simplifies the design, allowing the hydrogen separator made of silica membranes to perform at its optimal temperature of 600°C, while the catalytic H_2S decomposition proceeds in the CRs at about 900°C.

The conventional membrane architecture (CA), the one where the catalyst is placed within the wall of the membrane, is the most studied on the laboratory scale and most frequently reported in the literature. However, the application of the conventional architecture to H_2S decomposition is limited by the current ceramic membrane thermal stability [72] and a mismatch between the membrane performance, in terms of flow through and the heat flux, which could be applied to the catalytic tubes. Such imbalance would, from an engineering point of view, require a large and impractical heat transfer surface.

The performance of a single OA section was simulated using Excel Solver for the 10% MoS_2/Al_2O_3 catalyst developed at UNISA and a ceramic hydrogen permselective membrane developed at CanmetENERGY and described recently by Galuszka and Giddings [72]. All calculations assumed thermodynamically

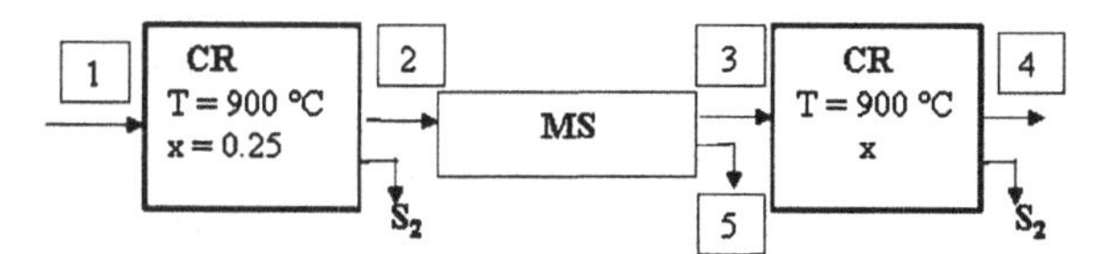

Fig. 8.6 Schematic of a single section of the OA

allowable conversions; the five stream compositions shown in Fig. 8.6 were considered.

The feed stream containing 10% of H_2S in nitrogen enters the first CR operated at 900°C, where 25% of H_2S is converted to hydrogen and sulphur as thermodynamically allowed and experimentally confirmed at UNISA laboratories. Sulphur is condensed at the exit of the first CR and the remaining products are fed to the tube side of the MS where part of hydrogen is removed through the membrane and recovered as a hydrogen-rich stream. The hydrogen- and sulphur-depleted stream containing unreacted H_2S, N_2 and remaining H_2, is fed to the second CR reactor.

The stream exiting the MS and entering the second CR may contain between 0 and 2.5% of hydrogen, depending on the efficiency of the MS module. If all H_2 produced in the first CR is removed in the MS, then the plant containing the two CRs and one MS, will give a net combined H_2S conversion of 43.75%. It needs to be noted that the sulphur removal alone after the first CR contributes to the overall conversion increase by approximately 5%.

The effectiveness of the MS module depends on H_2 membrane permeance, separation and pressure difference across the membrane (ΔP). Consequently, to keep ΔP at its maximum value, the partial pressure of hydrogen on the outside of the membrane should be as low as possible. Hydrogen would, therefore, be swept from the outside surface by employing a flow of a non-reactive gas, such as steam or be pumped away. For a fixed (ΔP), the dependence of the amount of hydrogen removed by the MS membrane and the total achievable H_2S conversion on H_2 permeance were calculated and are shown in Figs. 8.7 and 8.8.

A catalyst placement inside the MS module working at 600°C could be considered. At this temperature, an equilibrium conversion for the H_2S decomposition reaction is about 3%. If equilibrium is reached inside the MS containing the catalyst, then an additional shift towards the H_2S decomposition products will be expected.

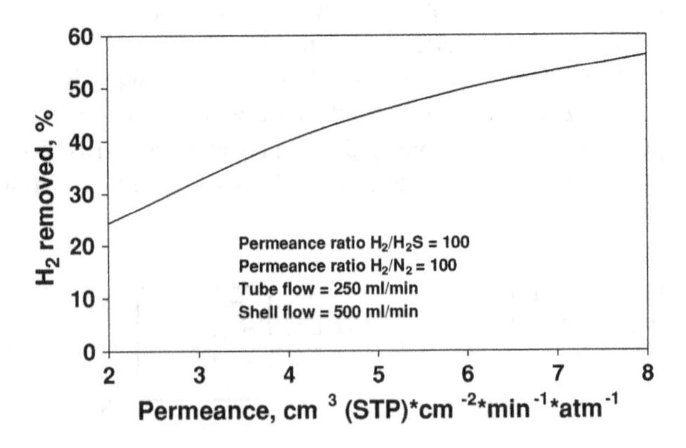

Fig. 8.7 H_2 amount removed by the MS as a function of the membrane H_2 permeance

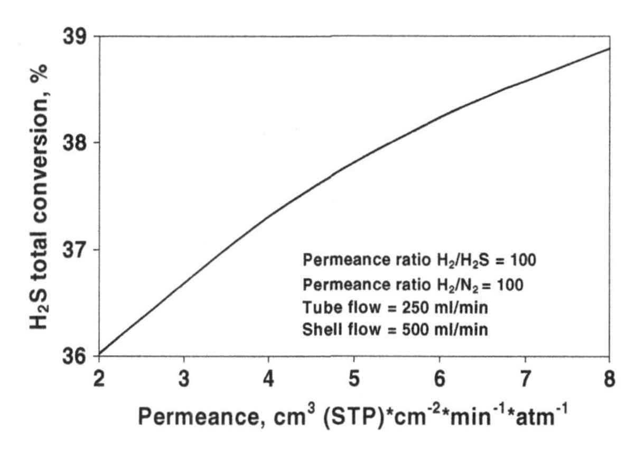

Fig. 8.8 H_2S total conversion as a function of the H_2 permeance of the MS

8.6 Practical Realization: Preliminary Technical Analysis and Economics

The most important element for the practical realization of the proposed membrane-assisted H_2S decomposition process is the use of the Claus process to generate heat for the thermal decomposition of H_2S and the use of a lower temperature thermal decomposition, coupled with permeable membrane separation to allow lower temperatures to be employed throughout. Residual H_2S would be disposed of during the Claus process; its heat content would be used as a partial or complete energy source for the process according to the following reaction:

$$H_2S + \text{heat} \leftrightarrow H_2 + \frac{1}{2}S_2 \tag{8.2}$$

$$10H_2S + 5O_2 \leftrightarrow 2H_2S + SO_2 + \frac{7}{2}S_2 + 8H_2O + \text{heat} \tag{8.3}$$

Consequently, relatively inexpensive materials may be used for the apparatus, and there would normally be no requirement for external energy and no generation of carbon dioxide.

The decomposition gas leaving the reactor would normally be quenched, i.e. quickly reduced in temperature, to avoid back reaction of hydrogen and sulphur in the decomposition gas, which may take place if the gas is allowed to cool slowly. This can be achieved by passing the decomposition gas through a suitable heat exchanger.

Before ultimately burning the residual H_2S in the thermal step, the decomposition gas containing unreacted H_2S would be heated at least one more time to the decomposition temperature of up to about 1000°C by heat generated in the thermal step, followed each time by further separation of hydrogen in the MS module.

A three-step CRMS configuration for the novel process is shown in Fig. 8.9.

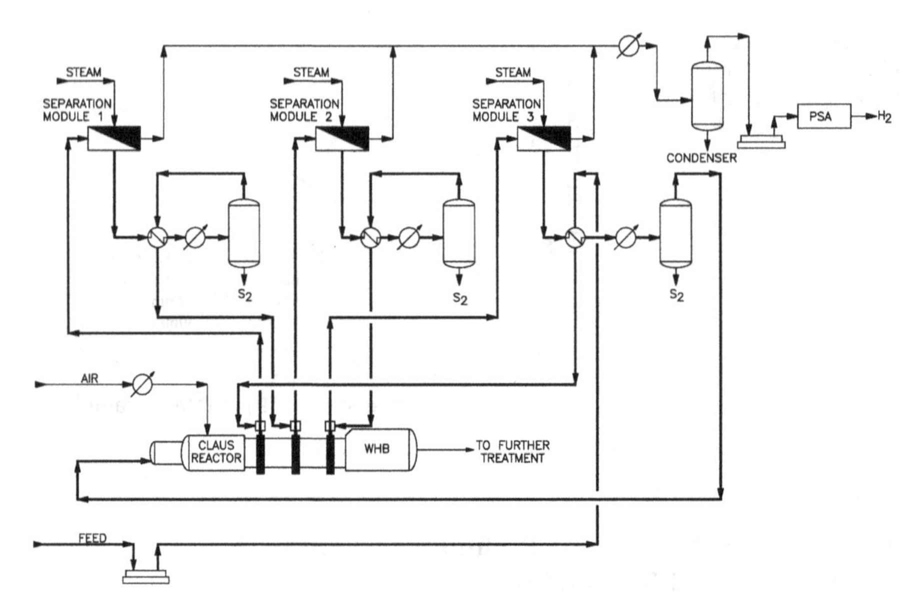

Fig. 8.9 Three-step CRMS configuration for the novel H_2S decomposition process

A stream containing 75% H_2S would be compressed to 8 atm, preheated to 550°C downstream of the third module, before entering the first reaction step consisting of fixed bed catalytic tubes. These CR tubes are immersed in the Claus reaction chamber, where heat is exchanged with the Claus gases and the H_2S stream to provide the energy required to carry out the H_2S decomposition reaction.

The proposed H_2S decomposition catalytic CR has a non-conventional, the so-called regenerative geometry. The gaseous products of the H_2S decomposition leave the bottom of the CR at 960°C through an internal riser, as shown in Fig. 8.10. This configuration allows for heat exchange between the decomposition gases and fresh stream, reducing the temperature of the product stream to 600°C and the overall heat duty of the process.

The product stream, containing H_2S, H_2, and S_2, leaves through the top of the CR and is at the right temperature to enter the first MS module directly. Hydrogen is removed in the MS module, and the retentate is cooled to the dew point temperature to separate sulphur; it enters the second CRMS at 550°C. Again, the decomposition gas leaves the second CR at 600°C, enters the second MS where hydrogen is removed, is cooled down to separate sulphur and recycled to the third and final reaction step. The decomposition gas leaving the third CR at 600°C, is then routed to the third MS to remove hydrogen, cooled down to separate sulphur and recycled to the Claus reactor to treat the unconverted H_2S and produce the required reaction heat through the Claus process.

Operating conditions, temperature, pressure, removal rate etc., of the three CRMS units are set to generate the partial pressure of hydrogen high enough to facilitate an optimal H_2 separation through a membrane and the H_2S conversion,

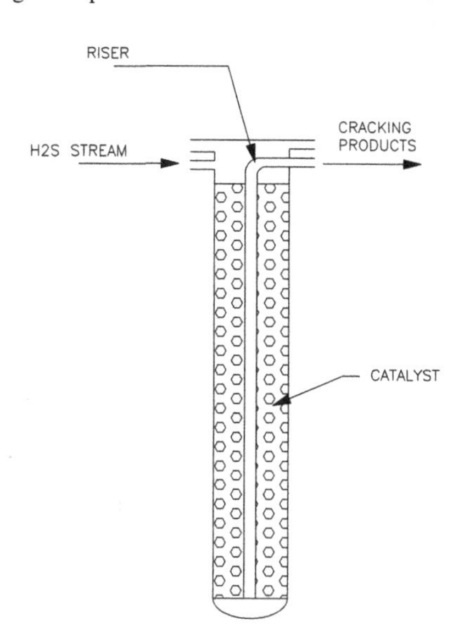

Fig. 8.10 Regenerative geometry of the tubular CR

Table 8.1 Main process parameters used for techno-economical comparison

Case	Conventional unit/SMR	Membrane-assisted
Overall sulphur production, ton/day	300	300
H_2 production, Nm^3/h	2200[a]	2200
Natural gas consumption, Nm^3/h	880[b]	–
CO_2 emission, ton/day	41.4	–[c]
Duty cracking, MM cal/h	–	3.0[d]
MP steam, ton/h	–	6.0
Power consumption, kWh/h	–	650[e]

[a] Based on 90% H_2 removal at each membrane step
[b] Based on 0.4 Nm^3 net gas per Nm^3 of H_2
[c] No CO_2 emission associated with steam production
[d] Based on conversion of 130 kmol/h of H_2S at 7/6/5 atm and 500/960°C
[e] It includes power for compressing the H_2S stream, the H_2 product, deducting 1/3 of the air compressor from the Claus unit and methane compressing in the SMR

which is sustainable with the heat and the temperature generated in the Claus reactor.

Hydrogen streams are cooled, compressed and further cooled before a final purification with a dedicated Pressure Swing Adsorber (PSA) to correct less than ideal hydrogen selectivity of silica membranes. The PSA polisher would be able to achieve hydrogen purity in excess of 99.999%.

In the proposed scheme, the heat used for the decomposition process in the Claus reactor reduces the amount of medium pressure steam available at battery limits. However, since no natural gas is used for hydrogen production, the process does not emit any CO_2.

Table 8.2 Key economic parameters

Natural gas price	0.22 €/Nm^3
MP steam	10 €/ton
Electricity price	0.07 €/kWh
CO_2 penalization	40 €/ton
Depreciation	10%/year of investment
Maintenance costs	2.5% of the investment
Maintenance membrane	Replacement every 4 years
Operating h/year	8,400
Investment cost (300 ton/day Sulphur)	110 MM€[a]
Investment cost (200 ton/day Sulphur)	85 MM€[b]
Modification to the Claus unit	21.25 MM€
Investment cost (2200 Nm^3/h H_2 plant)	8.0 MM€

[a] Referred single train with tail gas treatment (TGT) and overall sulphur removal of 99.9%, TKT basis
[b] Referred single train with PSA, H_2 purity 99.999%, TKT basis

Main process parameters used for the techno-economical comparison are summarized in Table 8.1, where the novel process is compared with a conventional Claus unit + tail gas treatment (TGT) and steam methane reforming (SMR) for hydrogen production.

Key economic parameters used to evaluate the novel process are shown in Table 8.2. A simple approach, based on the overall operating cost, was used to compare the conventional with the proposed CRMS technology.

The investment cost was estimated using a standard estimating tool and Tecnimont KT's long-standing experience with building Claus sulphur units and hydrogen plants.

For the membrane-assisted process evaluation, a 200 ton/day unit was considered together with the costs to modify the Claus reaction chamber by inserting the catalytic tubes. The cost of separation modules, vessels, exchangers and rotating machinery was also added. Table 8.3 reported the estimated equipment costs. For the overall investment cost, a 250% or 2.5 multiplying factor was used, which is commonly adopted for similar estimates.

8.6.1 Variable Operating Costs (VOC)

The proposed process is penalized by the extra power consumed and the loss of a part of MP steam, when compared to the conventional process of natural gas consumption and associated CO_2 emissions. Table 8.4 shows a comparison of the extra variable operating costs (VOC) for both cases. Owing to the impact of natural gas costs, the difference between the two schemes is considerable, more than 1.3 MM€ in favour of the membrane-assisted scheme.

Table 8.3 Estimated cost of equipment for the membrane-assisted scheme (MM€)

Modification to Claus unit including catalytic tubes[a]	2.00
Rotating machinery[b]	2.00
Membrane modules[c]	2.7
Heat exchangers and vessel	1.30
PSA and other	0.50
Total	8.5

[a] Based on 48 tubes, 4″ ID and 4 m long, based on 50000 kcal/m^3 as heat flux
[b] Based on investment costs of 2000 per kW inst
[c] Overall estimated membrane surface, 1800 m^2 at a cost of 1500 €/m^2

Table 8.4 VOC comparison on annual basis, for conventional SMR versus proposed process, MM€

Utilities	Conventional + SMR	Membrane-assisted
Natural gas (0.22 €/Nm^3)	1.63	–
Power (0.07 €/kWh)	–	0.38
MP steam (10 €/ton)	–	0.50
CO_2 emissions (40 €/ton)	0.57	–
Total	2.20	0.88

8.6.2 Overall Operation Costs

In Table 8.5, the overall investment costs and depreciation were reported. At this early stage, modifications required to the Claus unit are not completely mastered; however, a sizable difference between the two schemes emerges, and the novel scheme has a slightly better yearly depreciation.

Table 8.6 presents the overall change in operating costs. For the O&M of the MS modules, complete replacement once every 4 years was considered. A predicted difference of 2.175 MM€/year in operational costs, which represents a 13% reduction, is an appealing figure, considering that of the O&M costs, the membrane replacement cost comprises a substantial portion.

Table 8.5 Comparison of the yearly depreciation on annual basis, for the conventional SMR and the new proposed process

Technology	H_2 plant MM€	Claus unit MM€	Total investment MM€	Yearly depreciation MM€
Conventional	8	110	118	11.8
Membrane-assisted	–	85 + 21.25	106.3	10.63
Total difference, MM€	8	3.75	11.75	1.17

Table 8.6 Comparison of overall operating costs on annual basis for the conventional SMR and new proposed process

Technology	VOC MM€	Yearly depreciation MM€	O&M MM€	Total
Conventional	+2.20	11.80	2.95	16.95
Membrane-assisted	+0.88	10.63	2.59 + 0.675	14.10
Overall difference	+1.32	+1.17	−0.315	+2.175

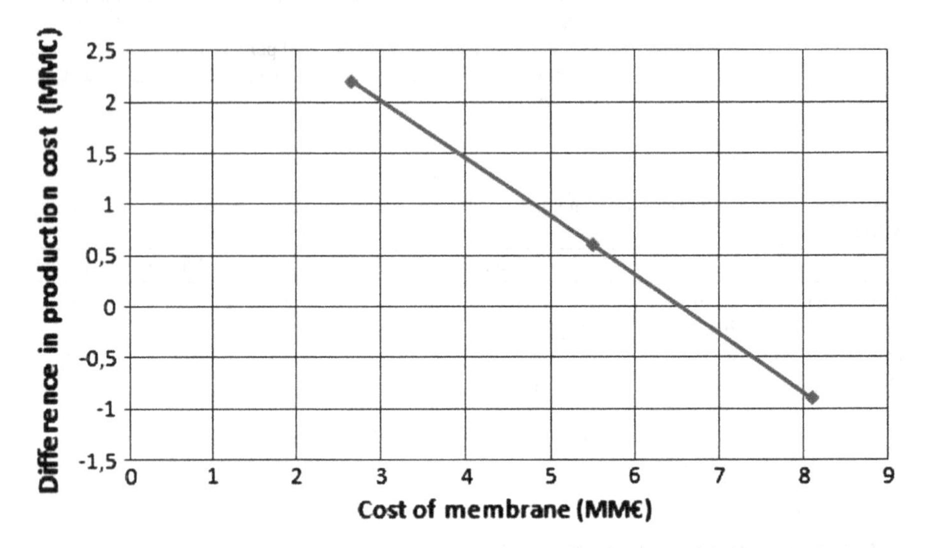

Fig. 8.11 Difference of H_2 production costs between the conventional and CRMS-based process versus cost of membrane

Sensitivity analysis using the cost of membrane as parameter is shown in Fig. 8.11. To nullify the operating cost difference, the membrane modules' cost must rise significantly to a unit cost of almost 6500 €/m^2.

It is likely that advances in membrane technology and production will further reduce the above figures, thereby increasing the differential production cost and making the novel process even more attractive.

Acknowledgments We wish to acknowledge the financial support provided by the Canadian Federal Government Program on ecoEnergy Technology Initiative, and by Tecnimont KT S.p.A. The participation of Mr. Terry Giddings of CanmetENERGY in Ottawa, Canada, is also appreciated.

References

1. Cox BG, Clarke PF, Pruden BB (1988) Economics of thermal dissociation of H_2S to produce hydrogen. Int J Hydrogen Energy 23:531–544
2. CAPP Report (2007) Oil sands: benefits to Alberta and Canada, today and tomorrow, through a fair, stable and competitive fiscal regime

3. Raymont MED (1974) Hydrogen sulfide thermal decomposition. Ph.D. Thesis, University of Calgary, Calgary, AB, Canada
4. Kaloidas VE, Papayannakos NG (1989) Kinetics of thermal non-catalytic decomposition of hydrogen sulphide. Chem Eng Sci 44(11):2493–2500
5. Fletcher EA, Noring JE, Murray JP (1984) Hydrogen sulfide as a source of hydrogen. Int J Hydrogen Energy 9(7):587–593
6. Noring JE, Fletcher EA (1982) High temperature solar thermochemical processing—hydrogen and sulfur from hydrogen sulfide. Energy 7(8):651–666
7. Yang BL, Kung HH (1994) Hydrogen recovery from hydrogen sulfide by oxidation and by decomposition. Ind Eng Chem Res 33(5):1090–1097
8. Chivers T, Hyne JB, Lau C (1980) The thermal decomposition of hydrogen sulfide over transition metal sulfides. Int J Hydrogen Energy 5(5):499–506
9. Gregory TD, Feke DL, Angus JC, Brosilow CB, Landau U (1980) Electrolysis of liquid hydrogen sulphide. J Applied Electrochemistry 10(3):405–408
10. Al-Shamma LM, Naman SA (1990) The production and separation of hydrogen and sulfur from thermal decomposition of hydrogen sulphide over vanadium oxide/sulphide catalysts. Int J Hydrogen Energy 15(l):l–15
11. Luinstra EA (1995) Hydrogen from H_2S: technologies and economics. Sulfotech Research, CA
12. Raymont MED (1975) Make hydrogen from hydrogen sulphide. Hydrocarb Process 54:139–142
13. Chivers T, Lau C (1985) The thermal decomposition of hydrogen sulfide over alkali metal sulfides and polysulfides. Int J Hydrogen Energy 10(1):21–25
14. Plummer MA (1994) Process for recovering sulfur and hydrogen from hydrogen sulfide. U.S. Patent 5,334,363
15. Plummer Mark A, Cowle Scott W (2006) Chemical mechanisms in hydrogen sulfide decomposition to hydrogen and sulfur. Mol Simul 32(2):101–108
16. Borgarello E, Kalyanasundaram K, Gratzel M Pelizzetti E (1982) Visible light Induced generation of hydrogen from H_2S in CdS-dispersions, Hole transfer catalysis by RuO_2. Helv Chim Acta 65:243–248
17. Borgarello E, Serpone N, Gratzel M, Pelizzetti E (1986) Photodecomposition of H_2S in aqueous alkaline media catalyzed by RuO_2-loaded alumina in the presence of cadmium sulfide. Application of the inter-particle electron transfer mechanism. Inorg Chim Acta 112(2):197
18. Kalyanasundaram K, Borgarello E, Gratzel M (1981) Visible light induced water cleavage in CdS dispersions loaded with Pt and RuO_2 hole scavenging by RuO_2. Helv Chim Acta 64:362–366
19. Kazuyuki T, Hideyuki T, Takatoshi M (2007) Materia 46(3):162–165
20. Huang CP, Linkous CA (2007) UV photochemical option for closed cycle decomposition of hydrogen sulfide. US patent 7220390B1
21. Linkous CA, Huang CJ, Fowler R (2004) UV photochemical oxidation of aqueous sodium sulfide to produce hydrogen and sulphur. J Photochem Photobiol A 168:153–160
22. Ma G, Yan H, Shi J, Zong X, Lei Z, Li C (2008) Direct splitting of H_2S into H_2 and S on CdS-based photocatalyst under visible light irradiation. J Catal 260:134–140
23. Zhang L, Wang Y, Bai X (2008) Photocatalytic decomposition of hydrogen sulfide to produce hydrogen over CdS/ZnO composite photocatalysts. Huaxue Yu Nianhe 30(6):5–8, 12
24. Xu H, Fu X, Bai X (2008) UV light catalytic decomposition of hydrogen sulfide to produce hydrogen. Huaxue Yu Nianhe 30(4):9–12
25. Argyle MD, Ackerman JF, Muknahallipatna S, Hamann JC, Legowski S, Zhang J, Zhao G, Alcanzare RJ, Wang L, Plumb OA (2004) Novel composite hydrogen-permeable membranes for non-thermal plasma reactors for the decomposition of hydrogen sulfide. DE-FC26-03NT41963
26. Thomas JR (1997) Particle size effect in microwave-enhanced catalysis. Catal Lett 49:137

27. Subrahmanyam CH, Renken A, Kiwi-Minsker L (2008) Non-thermal plasma catalytic reactor for hydrogen production by direct decomposition of H_2S. Optoelectro Nanomater 10(8): 1991–1993
28. Bolmer PW (1966) US Patent 3,249,522
29. Bolmer PW (1968) US Patent 3,409,520
30. Johnson GC (1966) US Patent 3,266,941
31. Kalina DW, Mass ET Jr (1985) Indirect hydrogen sulfide conversion-I. An acidic electrochemical process. Int J Hydrogen Energy 10(3):157–162
32. Mizuta S, Kondo W, Fujii K, Iida H, Isshiki S, Noguchi H, Kikuchi T, Sue H, Sakai K (1991) Hydrogen production from hydrogen sulfide by the Fe–Cl hybrid process. Ind Eng Chem Res 30:1601–1608
33. Mbah J, Krakow B, Stefanakos E, Wolan J (2008) Electrolytic splitting of H_2S using $CsHSO_4$ membrane. J Electrochem Soc 155(11):E166–E170
34. Edlund DJ, Frost CB, Pledger JR, Reynolds TA, Babcock WC (1995) A catalytic membrane reactor for facilitating the water-gas-shift reaction at high temperatures—phase II. Final Report to the U.S. Department of Energy on Contract No. DE-FG03-91-ER81229, Bend Research Inc., Bend, Oregon
35. Edlun D (1996) A membrane reactor for H_2S decomposition. DOE/ER/81419-97/C0749
36. Zaman J, Chakma A (1995) A simulation study on the thermal decomposition of hydrogen sulphide in a membrane reactor. Int J Hydrogen Energy 20(1):21–28
37. Morreale BD, Ciocco MV, Howard BH, Killmeyer RP, Cugini AV, Enick RM (2004) Effect of hydrogen-sulfide on the hydrogen permeance of palladium-copper alloys at elevated temperatures. J Membr Sci 241:219–224
38. Osemwengie UI, Morreale BM, Killmeyer RP, Enick RM, Howard BH (2006) Performance of Pd-alloy membranes for hydrogen separation from mixed feed streams containing 1000 ppm H_2S. Abstract, AIChE 2006 Spring National Meeting, Orlando
39. Pomerantz N, Ma YH (2007) Effect of H_2S poisoning of Pd/Cu membranes on H_2 permeance and membrane morphology. Am Chem Soc DC Coden: 69JNR2 Conference. AN 2007:882084
40. Howard B, Rothenberger K, Killmeyer R, Enik R, Cugini A (2003) The hydrogen permeability and sulphur resistance of palladium-copper alloys at elevated temperature and pressure. Mater Res Soc Symp Proc 752:277–282
41. Aboud S, Ozdogan E, Wilcox J (2009) Ab initio studies of palladium-niobium alloys for hydrogen separation. Abstracts of Papers, 237th ACS national meeting, Salt Lake City, UT, USA, 22–26 March 2009
42. Koros WJ, Fleming GK (1993) Membrane-based gas separation. J Membr Sci 83:1–80
43. Freeman BD, Pinnau I (2004) Gas and liquid separations using membranes: an overview. In: Pinnau I, Freeman BD (eds) Advanced materials for membrane separations. ACS symposium series 876, American Chemical Society, Washington, DC, pp 1–21
44. Roa F, Way JD (2003) Influence of alloy composition and membrane fabrication on the pressure dependence of the hydrogen flux of palladium copper membranes. Ind Eng Chem Res 42:5827–5835
45. Lee D, Zhang L, Oyamaa ST, Niuc S, Saraf RF (2004) Synthesis, characterization, and gas permeation properties of a hydrogen permeable silica membrane supported on porous alumina. J Membr Sci 231:117–126
46. Trujillo FJ, Hardiman KM, Adesina AA (2008) Catalytic decomposition of H_2S in a double-pipe packed bed membrane reactor: numerical simulation studies. Chem Eng J 143:273–281
47. Dokiya M, Kameyama T, Fukuda K (1978) Jpn Patent 78,130,291
48. Toyobo Co. Ltd (1980) Jpn Patent 80,119,439
49. Abe F (1987) Eur Pat Appl 228, 885
50. Gavalas GR, Megiris CE (1990) Synthesis of SiO_2 membrane on porous support and method of the same. US Patent 4902307
51. Peachey NM, Dye RC, Show RC, Birdsell SA (1998) Composite metal membrane. US Patent 5738708

52. Blach Vizoso R (2002) Catalytic membrane reactor that is used for the decomposition of hydrogen sulphide into hydrogen and sulphur and the separation of the products of said decomposition. US Patent 2004141910 (a1)
53. Agarwal PK, Ackerman J (2006) Membrane for hydrogen recovery from streams containing hydrogen sulfide, University of Wyoming, USA. US Patent Application Publication 5 pp
54. Nishizawa T, Tanaka Y, Hirota K (1979) Decomposition of hydrogen sulfide and enrichment of hydrogen produced by use of thermal diffusion columns. Int Chem Eng 19:517
55. Chivers T, Lau C (1987) The use of thermal diffusion columns reactors for the production of hydrogen and sulfur from the thermal decomposition of hydrogen sulfide over transition metal sulfides. Int J Hydrogen Energy 12(8):561–569
56. Hirota K (1977) Thermal decomposition of hydrogen sulfide in gas mixtures. Japan. Kokai 17(52):173
57. Chivers T, Lau C (1987) The thermal decomposition of hydrogen sulfide over vanadium and molybdenum sulfides and mixed sulfide catalysts in quartz and thermal diffusion column reactors. Int J Hydrogen Energy 12(4): 235–243
58. Faraji F, Safarika I, Strausz OP, Yildirimb E, Torresc ME (1998) The direct conversion of hydrogen sulfide to hydrogen and sulfur. Int J Hydrogen Energy 23:451–456
59. Dowling NI, Hyne JB, Brown DM (1990) Kinetics of the reaction between hydrogen and sulfur under high-temperature Claus furnace conditions. Ind Eng Chem Res 29(12):2327
60. Darwent de B, Roberts R (1953) Proc Roy Soc (Lond) A 216: 344
61. Hawboldt KA, Monnery WD, Svrcek WY (2000) New experimental data and kinetic rate expression for H_2S pyrolysis and re-association. Chem Eng Sci 55:957–966
62. Fukuda K, Doklya M, Kameyama T, Kotera Y (1978) Catalytic decomposition of hydrogen sulfide. Ind Eng Chem Fundam 17:4
63. Kaloidas VE, Papayannakos NG (1991) Kinetic studies on the catalytic decomposition of hydrogen sulfide in a tubular reactor. Ind Eng Chem Res 30(2):345
64. Monnery WD, Hawboldt KA, Pollock A, Svrcek WY (2000) New experimental data and kinetic rate expression for H_2S pyrolysis and re-association. Chem Eng Sci 55:957–966
65. Kappauf T, Fletcher EA (1989) Hydrogen and sulfur from hydrogen sulfide VI. Solar thermolysis. Energy 14:443–449
66. Steinfeld A (2005) Solar thermochemical production of hydrogen: a review. J Solar Energy 78:603–615
67. Zhaoa G, Sanil J, Zhanga J, Hamannb JC, Muknahallipatnab SS, Legowskib S, Ackermana JF, Argylea MD (2007) Production of hydrogen and sulfur from hydrogen sulfide in a nonthermal-plasma pulsed corona discharge reactor. Chem Eng Sci 62:2216
68. Roussy G, Pearce JA (1995) Foundations and industrial applications of microwaves and radio frequency fields. Wiley, New York
69. Zhang X, Hayward DO (2006) Applications of microwave dielectric heating in environment-related heterogeneous gas-phase catalytic system. Inorg Chim Acta 359:3421–3433
70. Zhang X, Hayward DO, Mingos MP (1999) Apparent equilibrium shifts and hot-spot formation for catalytic reactions induced by microwave dielectric heating. Chem Commun 975–976
71. Kaloidas VE, Papayannakas NG (1987) Int J Hydrogen Energy 12:403
72. Galuszka J, Giddings T (2011) Silica membranes-preparation by chemical vapour deposition and characteristics. In: Basile A, Gallucci F (eds) Membranes for membrane reactors: preparation, optimization and selection, Chap 12. Wiley, New York (in press)
73. Kameyama T, Dokiya M, Fujishige M, Yokokawa H, Fukuda K (1981) Possibility for effective production of hydrogen from hydrogen sulphide by means of a porous vycor glass membrane. Ind Eng Chem Fundam 20:97–99
74. Kameyama T, Dokiya M, Fujishige M, Yokokawa H, Fukuda K (1983) Production of hydrogen from hydrogen sulphide by means of selective diffusion membranes. Int J Hydrogen Energy 8:5–13

75. Ohashi H, Ohya H, Aihara M, Negeshi Y, Semenova SI (1998) Hydrogen production from hydrogen sulphide using membrane reactor integrated with porous membrane having thermal and corrosion resistance. J Membr Sci 146:39–52
76. Vizoso RB (2004) Catalytic membrane reactor for breaking down hydrogen sulphide into hydrogen and sulfur and separating the products of this breakdown. US Patent Application, US2004/0141910 A1
77. Akamatsu K, Nakane M, Sugawara T, Hattori T, Nakao S (2008) Development of a membrane reactor for decomposing hydrogen sulphide into hydrogen using a high-performance amorphous silica membrane. J Membr Sci 325:16–19
78. Edlund DJ, Friesen DT (1993) Hydrogen-permeable composite metal membrane and uses thereof. US Patent 5,217,5006
79. Edlund DJ, Pledger WA (1993) Thermolysis of hydrogen sulfide in a metal-membrane reactor. J Membr Sci 77:255–264
80. Edlund DJ, Pledger WA (1994) Catalytic platinum-platinum based membrane reactor for removal of H_2S from natural gas stream. J Membr Sci 94:111–119
81. Bartholomew CH, Agrawal PK, Katzer JR (1982) In: Eley DD, Pines H, Weisz PB (eds) Sulfur poisoning of metals, vol 31. Advances in Catalysis Academic, New York, pp 135–241
82. Kotera Y (1976) The thermochemical hydrogen program at N.C.L.I. Int J Hydrogen Energy 1:219–220
83. Zazhigalov VA, Gerei SV, Rubanik MYa (1975) Relationship in the catalytic reaction between hydrogen and sulfur in the presence of metal sulphide I. Kinet Katal 16(4):967–974

Chapter 9
Alkanes Dehydrogenation

Moshe Sheintuch and David S. A. Simakov

9.1 Process Description

Light alkenes (olefins) are among the most important intermediate products in chemical industry, e.g., ethylene is required for the production of polyethylenes and polyvinylchlorides. Nowadays, light olefins are commonly obtained by steam cracking (SC) and fluid catalytic cracking (FCC) of light oil fractions. For example, most propylene is produced as co-product in steam crackers (>55%) and as by-product in FCC units (~35%), while only small fraction (<10%) is produced by alternative technologies, such as propane dehydrogenation. For ethane dehydrogenation, ~84% selectivity and ~54% ethane conversion are commonly obtained by SC at 800°C. As both SC and FCC require high temperatures, coking and side reactions are among major drawbacks (SC and FCC are highly endothermic processes with extensive coke formation).

Catalytic dehydrogenation (DH) becomes a growing branch in petrochemical industry, as a route to obtain alkenes from low-cost feedstocks of saturated hydrocarbons. Catalytic dehydrogenation is currently a well-established commercial route for production of a number of important light olefins, e.g., propene. As compared to conventional cracking technologies, catalytic DH may provide better selectivity at lower temperatures, lowering also the coke deposition rate. In general, DH of paraffins is an endothermic equilibrium reaction, with conversions increasing with decreasing pressure and increasing temperature. Process temperature increases with decreasing carbon number to maintain conversion for a given pressure. The main side reaction that occurs during the DH process is cracking, which is primarily thermal in nature and results in the undesirable products and in coke formation. Another side reaction, which occurs during DH of isobutene, is the isomerization to the normal form.

M. Sheintuch (✉) and D. S. A. Simakov
Technion, Technion City, 32000 Haifa, Israel
e-mail: cermsll@technion.ac.il

M. De Falco et al. (eds.), *Membrane Reactors for Hydrogen Production Processes*,
DOI: 10.1007/978-0-85729-151-6_9,

Several examples of light alkanes DH are shown below:

$$C_2H_6 \rightarrow C_2H_4 + H_2,\ \Delta H^\circ_{298} = 137\,\mathrm{kJ/mol} \tag{9.1a}$$

$$C_3H_8 \rightarrow C_3H_6 + H_2,\ \Delta H^\circ_{298} = 124\,\mathrm{kJ/mol} \tag{9.1b}$$

$$iC_4H_{10} \rightarrow iC_4H_8 + H_2,\ \Delta H^\circ_{298} = 118\,\mathrm{kJ/mol} \tag{9.1c}$$

Temperatures as high as 900 and 750°C are required to achieve a conversion of 90% for ethane and propane at equilibrium, respectively (at atmospheric pressure). To obtain 70% conversion, the corresponding temperatures are 790 and 660°C.

9.1.1 Traditional Process Drawbacks

In general, DH of light alkanes (ethane, propane, butane, etc.) suffers from several common limitations: (i) thermodynamic restrictions on conversion, which imply high temperatures to obtain reasonable conversions; (ii) strong endothermicity, which requires heat supply, and several approaches for that purpose are described in the following sections; (iii) side reactions like thermal cracking, isomerization, and butadiene production, which become important at high residence times; and (iv) coke formation due to the high temperatures. Coking is one of the major reasons for catalyst deactivation, which is extremely fast (<1 h in industrial applications). As a result, catalytic-bed reactors have to be regenerated every short period of time (tens of minutes), to burn off the deposited carbon.

An additional mechanism of deactivation, due to catalyst particle agglomeration (sintering), leads to reduction in active specific surface area and, as a result, catalyst activity. Obviously, such deactivated catalytic pellets cannot be regenerated just by combustion of hydrocarbons over the deactivated catalytic bed (like in case of coking); this type of deactivation is rather irreversible. Moreover, periodic regenerations of the catalytic bed (to burn off deposited coke) make the sintering deactivation even more severe.

Catalytic oxidative dehydrogenation (ODH) of light alkanes helps to overcome (at least partially) some inherent drawbacks of non-oxidative dehydrogenation. Oxidative dehydrogenation reduces the process endothermicity and shifts the equilibrium toward alkenes generation, offering better conversions and selectivity. However, ODH suffers from its own drawbacks: (i) catalyst selection, as catalyst should be active for both oxidation and DH; (ii) insufficient selectivity; (iii) loss of hydrogen, which is a valuable by-product (inherent feature of ODH); (iv) flammability of the reactive mixture; (v) heat removal; and (vi) reaction runaways. The catalyst also should withstand high temperatures if hot spots emerge in a fixed bed as a result of addition of oxygen. Introduction of an oxidant to the process obviously raises safety concerns: heat has to be removed and a control is necessary to prevent runaways.

9.1.2 Dehydrogenation Catalysts

Catalytic DH is normally performed in tubular packed-bed reactors (PBR). Alumina-supported chromia (Cr_2O_3/Al_2O_3) is the most common DH catalyst for commercial applications, due to its high activity and selectivity. In isobutane DH, conversions of $\sim$55% and selectivities higher than 90% are normally obtained in industrial plants at temperatures below 600°C and nearly atmospheric pressure; the catalyst is regenerated every few hours [1]. The activity of Cr-based DH catalysts is dependent on the Cr loading; it is also affected by the process conditions, mainly temperature and time-on-stream [1]. For small Cr loadings, the DH activity increases almost linearly with Cr loading, whereas for high loadings (5–10 wt% Cr) the activity remains similar or even decreases. Unfortunately, the γ-alumina support (which is widely employed for commercial applications) catalyzes the side reactions of cracking and coking, resulting in fast catalyst deactivation. For example, in the propane DH, propane conversions of $\sim$50% and selectivity to propylene of $\sim$90% are normally obtained in the beginning of operation (at $\sim$600°C and atmospheric pressure), but the conversion drops to $\sim$40% in just several hours of operation [2, 3]. The deactivation is mainly attributed to the formation of coke. A constant DH activity for a certain period of time may be achieved by gradual increase of temperature during the operation.

Other supports with different acidic characteristics and the addition of alkali or alkaline-earth metals and other promoters are used [2, 3]. For example, the addition of tin (Sn), which affects the amount of the oxidized chromium, results in a significant decrease of the amount of deposited coke. The Sn addition also improves selectivity, but, unfortunately, reduces catalytic activity [3]. Alternative catalysts are extensively investigated and some of them are already in commercial use (e.g., Pt-based).

Alumina-supported platinum-based catalysts (Pt/Al_2O_3) have also been successfully employed for light paraffins DH. The main drawback is again deactivation, which is related not only to coking, but also to sintering. Promoters that can suppress carbon deposition or/and provide a strong interaction between Pt particles and promoter or support are of great importance. The deactivation due to coke formation can be prevented by the addition of various promoters. Bimetallic Pt-based catalysts supported on α-alumina (rather than traditionally employed γ-alumina) with the addition of promoters, such as indium and tin, provide high activity, selectivity, and stability and much smaller coke formation (decreased by a factor of 30–60) during DH of butane [4]. In propane DH, conversions of 40–50% and selectivity to propylene of 85–90% can be achieved at 2 bar and 570–580°C, using these catalysts [4]. There are evidences that the addition of Ce and Zn promoters could improve thermal stability of $Pt/\gamma\text{-}Al_2O_3$, enhancing platinum particles ability to resist to agglomeration [5]. Ce and Zn promoters improve chemisorption of hydrogen, maintaining catalyst activity; Zn also promotes propylene desorption, decreasing the degree of coking, and increasing the selectivity of propylene in propane DH (99% propylene selectivity for 35–40% propane conversion at 580°C).

In isobutane dehydrogenation, the reaction rate is usually expressed as

$$r_{iC4} = \frac{k\left(P_{iC4} - (P_{i=C4}P_{H_2})/K_{eq}\right)}{D(P_{iC4}, P_{i=C4}, P_{H_2})} \tag{9.2}$$

where P_i is the partial pressure of a specie i. This expression accounts for the inhibition due to adsorption of products or reactants or due to coke precursors. The kinetics of isobutane dehydrogenation on pellets of 0.35% Pt/γ-Al_2O_3, with or without addition of Sn or In promoters, was investigated by Lyu et al. [6, 7]. At temperatures between 500 and 600°C the reaction was found to be kinetic-controlled and the rate followed (9.2) with

$$D = P_{i=C4} + k''(p_{H_2})^{1/2} + k'''C_{coke} \tag{9.3}$$

The inhibition due to the amount of coke deposited on the catalyst was determined in a transient regime, due to deactivation, and the rate of coke formation was found to be proportional to isobutene pressure and inhibited by increasing hydrogen pressure. In a certain domain the rate of coke formation and destruction followed

$$r_C = (k_1 P_{i=C4} - k_2 p_{H_2} C_{coke})/D \tag{9.4}$$

with D given by Eq. 9.3.

While these results were obtained with relatively large hydrogen pressures and small conversions, the suggested expression provides a good qualitative model for the effect of hydrogen pressure.

Catalysts based on other metals, such as gallium and vanadium oxides, can be also employed in DH processes [8, 9]. For example, silica-supported gallium oxide catalyst has been found to be moderately active, but quite selective in propane dehydrogenation (up to 80%) and results in much less coking, $\sim$1/10 of that using a silica-supported chromium oxide [8]. There is an extensive research aimed to find new DH catalysts that will perform well at moderate temperatures, suffer less from coke deposition and maintain catalytic activity for long periods of time without regeneration.

Some of the limitations of the DH reactions can be overcome by the catalytic ODH. Various oxidizing agents can be used for ODH, e.g., air, oxygen, sulfur compounds, and halogens. The introduction of an oxidant into the DH reaction mixture makes it less endothermic and, therefore, the process can be performed at much lower temperatures. This in turn reduces the extent of side reactions, such as thermal cracking. Moreover, the DH thermodynamic limitations can be overcome, since removal of hydrogen from the reactive mixture (by oxidizing H_2 to water) shifts the equilibrium toward formation of products (alkenes). Though ODH offers some advantages with regard to reducing side reactions and overcoming equilibrium restrictions, it offers a relatively low selectivity and suffers from problems of flammability of the reaction mixture and reaction runaways; therefore, issues such as heat removal and optimization of oxidant-to-feed ratio are crucial for this process.

In ODH of light alkanes, such as propane, butane and isobutane, vanadium-based catalysts are commonly employed. Various types of supports and promoters (V_2O_5/ZrO_2, VO_x/TiO_2, V–Mg–O, VO_x/γ-Al_2O_3 and many others) can be used [10]. Other catalysts (e.g., Ni-based, Mo-based) can also be employed [10–12]. Significant progress has been done in the search for catalysts suitable for ODH, especially of ethane and propane. In the case of ethane, as reported in the literature ethylene yields are comparable to those obtained by SC, or even much better. Selectivities of 40–80% at conversions of 40–60% are normally reported for a relatively moderate temperature range of ~400–600°C using both V-based and other catalysts [12]. Some catalysts demonstrate a superior performance of ethylene selectivities of 80–90% at ethane conversions of up to 80% (in the same temperature range). For other alkanes, the reported yields are still significantly lower that those obtained in conventional processes (for propane ODH, up to ~60–80% selectivity at ~40–50% conversion is obtainable [12]). For low vanadium loadings (~0–6 wt% V), the activity of the V-based catalysts is increased with V loading, however, the selectivity shows a maximum for the intermediate loadings (in the same range). For example, in ethane ODH, the maximum ethane conversion of ~80% was found for 6 wt% V loading (at 570°C), while the maximum in ethylene selectivity (60%) was obtained for 5 wt% V, at ethane conversion of only 20% [13].

9.1.3 Industrial Applications

A number of catalytic DH technologies have been commercialized over the last decade and are reviewed below. Among the main processes currently used in the industry are CATOFIN®, Ole-flex™, and STAR® [14]. The CATOFIN *dehydrogenation* process, a proven route for the production of isobutylene and propylene, uses chromia-alumina catalyst fixed-bed reactors in a cyclic reaction-regeneration operation with a regeneration cycle of 20–30 min. During the hydrocarbon processing step, the feed is vaporized and raised to reaction temperature by exchange with various process streams and using a heater. The off-gas, which is a hydrogen-rich, can be sent to a pressure swing adsorption (PSA) unit to purify the hydrogen. Since the reactor temperature drops significantly during the reaction step, auxiliary equipment is required for the regeneration step, which is necessary to prepare the off-line reactors for the next reaction step.

The STAR process is also performed in fixed beds, over noble metal (Pt) supported by zinc and calcium aluminate and impregnated with various metals, but it is an ODH process. The catalyst maintains its activity for DH and is stable in the presence of steam and oxygen. The STAR process was proven for several applications, among them ODH of propane to propylene and ODH of butanes.

The Ole-flex process is a catalytic DH, performed in a multiple-stage unit including four radial-flow reactors packed with Pt-based catalyst, with charge and inter-stage heaters and feed-effluent heat exchangers. The process also includes a selective hydrogenation unit to eliminate diolefins and acetylenes.

9.2 Membrane Integration

The use of a hydrogen selective membrane for equilibrium-limited processes, such as DH, is a logical choice. Successful implementation of membrane reactors has a potential of replacing the currently used periodically regenerated fixed-bed reactors. Continuous removal of hydrogen by membrane separation should increase the DH conversion for a given temperature, providing also hydrogen, which is a valuable by-product. This in turn will allow operation at lower temperatures, therefore, reducing the coke formation and side reactions. On the other hand, using of H_2 selective membrane may results in extremely low H_2 partial pressures in the reaction mixture, leading to low selectivities and even increasing deactivation. Therefore, introduction of H_2 separation membranes into DH process should be supported by detailed analysis and process optimization.

9.2.1 Membranes for Hydrogen Separation

Metallic membranes, (Pd–Ag) alloys, are typically used for separation of H_2, either as an unsupported foil or a supported thin film. In these membranes, the hydrogen transport is by adsorption and atomic dissociation on one side of the membrane, dissolution in the membrane, followed by diffusion, and finally desorption (on the other side). Due to the H_2 dissociation step, H_2 separation is driven by a trans-membrane difference of the square roots of the hydrogen partial pressures. The preparation technologies of both unsupported and supported Pd–Ag membranes are well developed and such membranes are commercially available. Since the membrane reformer performance is limited by separation capability, optimization of membrane permeability is one of the important issues.

In early works Pd *unsupported foil* membranes of thickness of $\sim$70 μm were employed [15], giving rather moderate hydrogen fluxes. These membranes were used in commercial hydrogen separation units. However, further decrease of the foil thickness was challenging, since membranes became too fragile for the conditions of high temperature catalytic processes.

The search for a high permeability at low cost have led to the development of membranes composed of an ultrathin Pd–Ag layer (<10 μm) on a ceramic or stainless steel porous supports [16–18]. They combine the infinite selectivity to hydrogen of the dense Pd–Ag film with a mechanical strength of the porous support, which resistance to hydrogen transfer is negligible. These membranes are very promising, since they exhibit remarkably high fluxes [0.5–2.5 mol/(m^2 s), which is much higher than that of Pd–Ag foil membranes] and require relatively small quantities of Pd for their manufacturing.

Pinholes formation that can occur during the membrane fabrication or during the operation is still one of the major drawbacks. Membrane poisoning may be caused by strong adsorption of certain compounds (e.g., CO) and by coke deposition on the membrane surface, which is particularly relevant for DH processes. These shortcomings may lead to long-term hydrogen flux degradation and even to the membrane failure. There is still not enough information to quantitatively describe these effects.

Extensive research for substitutes for the expensive Pd membrane has been conducted. Porous membranes that consist of a highly porous metal or ceramic support with a thin top layer, tailored to have the desired selectivity, yield quite a high permeability but a relatively low selectivity. Some of the applications that have been tested on porous silica, vycor, alumina, and other membranes are listed in Saracco and Specchia [19] and Hsieh [20]. Most of the studies focused on selective permeation of products or reactants (mostly H_2, in some cases O_2) but the selectivity, which is determined by Knudsen diffusion, was very modest. While some improvement may be gained in ceramic membrane reactor when compared to conventional reactors; it is often attributed to the dilution effect of the sweep gas [21].

Ceramic hydrogen separation membranes that offer good selectivity to H_2 and have a potential of low-cost production may be very attractive for DH applications. Non-metallic porous hydrogen separation membranes like carbon [22–24], silica [25–27], and zeolitic [28, 29] are a subject of extensive investigation. These membranes typically contain pores of molecular dimensions. As a result, steric and energetic effects associated with the proximity of the pore wall, play an important role in transport. Transport mechanisms through membrane pores vary with the membrane pore size, the molecule size, and with the chemical interaction between the transported species and the membrane material. This diffusion regime is not well characterized and requires new computational tools like molecular dynamics. In addition to moderate selectivity to H_2, there are still many drawbacks in the preparation technology of ceramic and carbon membranes and they often fail to provide satisfactory durability under real operation conditions.

Carbon membranes are manufactured in a hollow fiber shape (Fig. 9.1) that can be arranged in multi-fiber modules. They exhibit moderate selectivity to H_2, e.g., H_2-to-isobutene permeability ratio of $\sim$100 may be achieved [22]. One of the main disadvantages of carbon membranes is their incompatibility for ODH, where DH and oxidation are directly coupled, i.e., carried out over the same catalytic bed.

Silica membranes are typically manufactured by depositing a thin film of silica on a suitable support (e.g., alumina) (Fig. 9.2). Two main routes for preparation of silica films are chemical vapor deposition (CVD) and sol–gel [26]. To provide resistance to water vapor, the membranes are typically doped by various metal additives (e.g. Zr and Ni). Silica membranes exhibit moderate selectivities to hydrogen, e.g., the H_2-to-N_2 selectivities are in the range of 100–500. Some membranes prepared by CVD

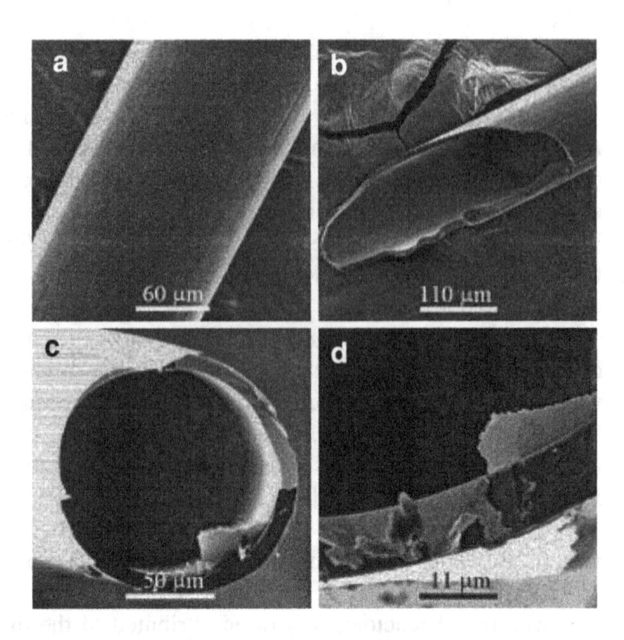

Fig. 9.1 SEM micrographs of the carbon membrane fibers (after [22])

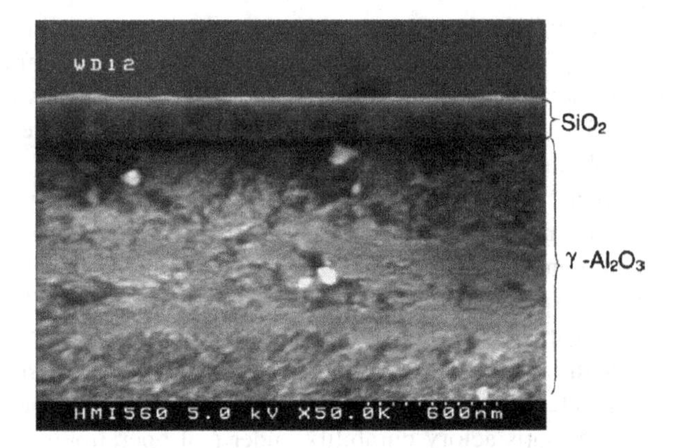

Fig. 9.2 Cross-section of the SiO_2 –coated alumina support with a 160-nm-thick SiO_2 layer (after [27])

methods may have quite high selectivities (e.g., >24,000 for H_2-to-CH_4 and >1,200 for H_2-to-CO_2), but provide low fluxes [26].

Zeolites are crystalline nanoporous materials with uniform nanosized pores (<1 nm) (Fig. 9.3). Selective permeation in zeolite membranes is based on molecular sieving and selective adsorption. Zeolite membranes have drawn attention as suitable membranes for DH applications due to their high thermal and chemical stability. When supported (Fig. 9.3), zeolite-based membranes also offer excellent mechanical strength, which is an important feature for DH applications. The permeation of single compounds in zeolitic membranes depends on the kinetic diameter of the molecule and size selectivity and they exhibit moderate selectivities to hydrogen.

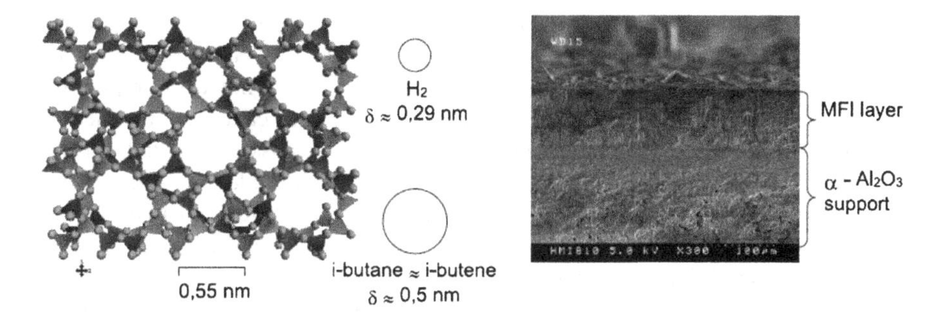

Fig. 9.3 Pore openings of an MFI-type zeolite membrane relative to the size of the H_2 and isobutane molecules (*left*) and SEM micrograph of the MFI-type membrane (*right*). Approximately 60-μm-thick zeolite layer is supported by an asymmetric α-Al_2O_3 carrier with 60 nm pores in the upper layer of the ceramic support (after [29])

9.3 Membrane Reactor Performance

Higher alkane DH conversions are expected in a membrane reactor equipped with a hydrogen selective membrane, when compared with a conventional packed-bed catalytic reactor. This, in turn, allows operating at lower temperatures, with expected slower catalyst deactivation. Laboratory membrane reactors for catalytic DH are commonly built as a concentric compartment packed-bed membrane reactor (PBMR). The hydrogen separation may be carried out from the membrane interior (the reactor shell is packed with a suitable catalyst), as well as from the reactor shell (in this case the membrane is packed with a catalyst). The separated hydrogen is normally swept by a sweep gas (e.g., N_2) to increase the transmembrane hydrogen pressure gradient and it may be conducted in co-current or countercurrent mode. Hydrogen is a valuable by-product, of value in the chemical industry and possibly for fuel cells that directly produce electricity in a pollution-free way with the efficiency much higher than that of heat engines.

9.3.1 Technology Drawbacks

Several difficulties arise when the hydrogen selective membrane is introduced into the reactor. The main drawbacks of this technology are related to the hydrogen separation effectiveness, membrane durability, and hydrogen separation-induced catalyst deactivation. The separation of hydrogen is temperature-activated but operation at high temperatures decreases the membrane durability and results in faster deactivation. Since DH equilibrium is favored by low pressures, the transmembrane pressure difference should be increased by the use of a sweep gas (e.g., N_2) rather than by increasing pressure, but then, in order to recover hydrogen from the H_2–N_2 mixture an additional PSA unit will be required.

Using air for sweeping separated hydrogen is much cheaper, since air is always available, but this approach suffers from the common to N_2-sweeping back-permeation problem and PSA is still required. It may be used, however, to provide the heat by oxidizing the separated hydrogen, which is catalyzed by the Pd membrane itself [30, 31]. This mode of operation is discussed in details in the following sections. Steam, which is also easily available in industrial application, can be used as a sweep gas, as well. In this case, the separation of hydrogen is straightforward, simply by cooling down the mixture to condense water.

Very little work has been devoted to thermal integration of endothermic DH with exothermic reactions (e.g., alkane oxidation). Such integration has a potential to improve significantly the overall efficiency of the process and it is discussed in Sect. 9.4 ("Future Perspectives"). In experimental works, the heat required for DH is typically supplied by electrical heaters, i.e., the reactor is operated under isothermal conditions.

9.3.2 Membrane Reactor Performance

Many studies of membrane reactors demonstrated conversions that surpassed that of equilibrium, at the same temperature and pressure, or that obtained in the same reactor, when operated as packed bed (i.e. with no sweep gas).

While performance of catalytic reactors is usually plotted vs. catalyst weight (W) to molar feed rare (F_0) ratio, the performance of membrane reactor depends on that parameter as well as on the membrane surface area. In fact in the limit of very fast reaction the performance depends only on the equilibrium coefficient and the ratio of the feed rate to the overall hydrogen separation rate. This ratio is sometimes referred to as a membrane Péclet number:

$$Pe_M = \frac{\text{feed rate}}{\text{separation rate}} \tag{9.5}$$

Effective H_2 separation is achieved for $Pe_M < 1$. For a given separation area, increasing the feed rate will diminish the conversion (e.g. [15, 23, 32, 33], Figs. 9.4, 9.5).

9.3.3 Palladium Membrane Reactors

A large number of hydrogenation and dehydrogenation reactions were tested in the early studies of dense-metal membrane reactors (see listing in Shu et al. [34], Hsieh [35], and Gryaznov and Orekhova [36]). Many works tested the dehydrogenation of cyclohexane to benzene as a model reaction since it can be carried out at low temperature with no side reactions and no deactivation; a conversion of 99.5% was achieved with a palladium membrane, compared with 18.7% at equilibrium, at 200°C [31].

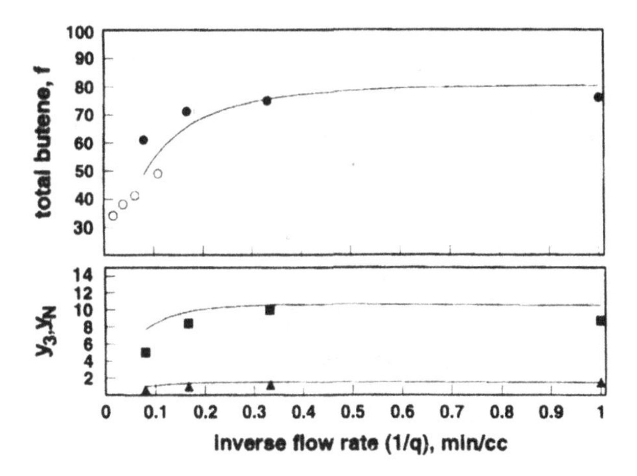

Fig. 9.4 Changes in total olefin yield (*f*), cracking (y_3, denoted by *triangles*), and isomerisation (y_N, *squares*) products with inverse flow rate during isobutane dehydrogenation in a Pd-tube packed with catalyst at 500°C (after [15])

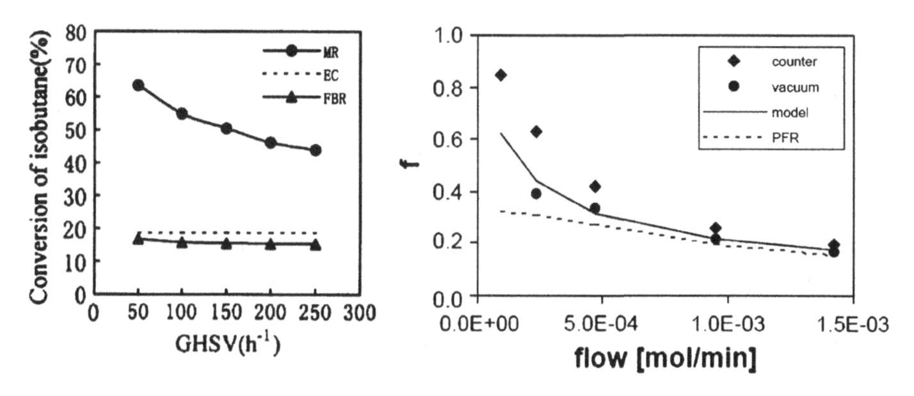

Fig. 9.5 *Left side* Pd–Ag membrane reactor isobutene conversion vs. feed space velocity, compared with equilibrium-limited and fixed-bed reactor (argon swept, $T = 723$ K, after [33]); *right side* carbon membrane reactor conversion, in the countercurrent sweep and vacuum modes, as a function of feed molar flows at 500°C; also denoted are the conventional (non-membrane) reactor conversion and the simulated countercurrent sweep mode behavior (after [23])

In early works, Pd and Pd alloys *unsupported foils* were used as a membrane [15, 31, 39], with a thickness mostly in the range of ~70–200 μm; thinner foils have been found to be too fragile for the DH operation conditions. Recently, great progress has been done in development of *supported thin film* Pd–Ag membranes, using various types of porous supports for this purpose; alumina [33] and stainless steel [37] are mostly employed. Due to their infinite selectivity to hydrogen, Pd–Ag membranes demonstrate superior performance, increasing greatly the DH conversions. Implementation of several DH processes in PBMRs has been studied, such as dehydrogenation of propane [15], butane [30], isobutane [15, 33, 37–39], and cyclohexane [31].

Sheintuch and Dessau [15] reported significant improvement in butane and propene yield in a PBMR equipped with Pd–Ru and Pd–Ag foils (250 and 76 μm,

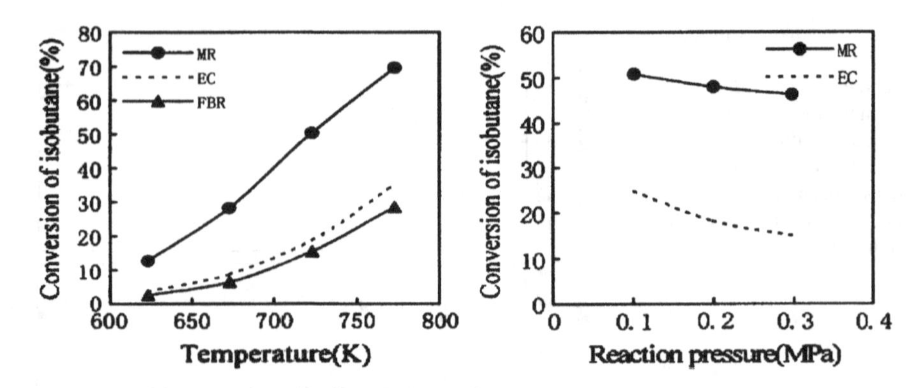

Fig. 9.6 *Left side* conversion of isobutene vs. reaction temperature at atmospheric pressure (after [33]); *right side* conversion of isobutene vs. reaction pressure (after [33])

respectively) packed with a Pt/Al_2O_3 catalyst, using N_2 as a sweep gas. The yields were 76% butane at 500°C, compared with 32% in equilibrium and 70% propene at 550°C vs. 23% in equilibrium (Fig. 9.4); relatively stable yields were achieved for periods of ~80 h (compared with <1 h in industrial applications). One of the important findings was the fact that for low feed rates (low Pe_M), when conversions are high and almost no H_2 is present in the reactive mixture, conversion could not be increased further by decreasing flow rate or by increasing temperature due to the fast deactivation in the absence of hydrogen. Also, high degrees of cracking and isomerization were found under these conditions (Fig. 9.4) due to the high residence times. Using the kinetics described in Eqs. 9.2–9.4, and treating C_{coke} as a coke-precursor whose coverage is the pseudo-steady-state solution of formation and destruction by reaction with hydrogen (Eq. 9.4), they were able to fit the experimental data and suggest optimal hydrogen pressure to be kept at the membrane.

The fabrication of a Pd–Ag alumina-supported membrane and its implementation for DH of isobutene has been reported by Guo et al. [33]. The membrane was composed of a ~9-μm-thick Pd–Ag film deposited by electroless plating on a 150-nm-thick porous α-alumina tube. The membrane exhibited high H_2 fluxes [~0.62–0.76 mol/(m^2 s) at 400–500°C] and isobutene conversions (over Cr_2O_3/Al_2O_3, sweeping by Ar) much higher than those in equilibrium (Fig. 9.6, *left side*). While increasing the temperature is favorable both for endothermic dehydrogenation and temperature-activated H_2 separation, pressure has opposite effects on the DH equilibrium and H_2 separation rate and the result of this contradiction can be seen in Fig. 9.6 (*right side*).

9.3.4 Deactivation

On the main concerns in DH technology is catalyst deactivation especially in the absence of H_2, as expected in membrane reactors. This may be prevented by

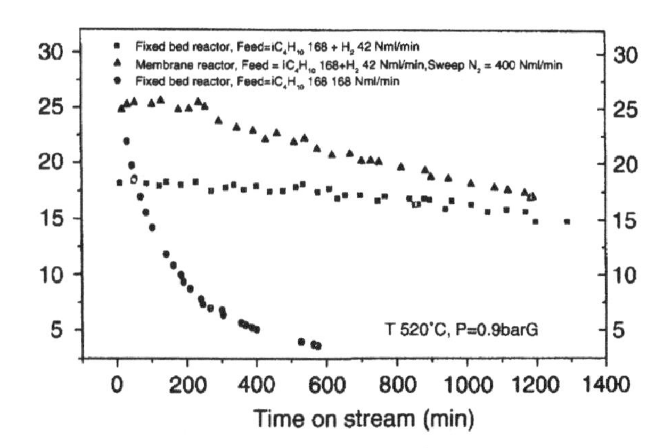

Fig. 9.7 Conversion of isobutane in a Pd–Ag PBMR over Pt/Al_2O_3 catalyst vs. time on stream; effect of hydrogen addition (after [37])

adjusting the operation parameters (e.g., the ratio of feed to sweep flow rate [23]) to avoid complete separation of hydrogen, or by addition of small fractions of hydrogen to the reactor feed. Consequently, the reactor should be designed considering the trade-off between the gain achieved by shifting the equilibrium and negative effects induced by faster catalyst deactivation at low hydrogen concentrations. Slower deactivation due to addition of H_2 to the reactor feed was demonstrated for both conventional and membrane reactors (Fig. 9.7). Quantitative description of these effects is required.

9.3.5 Ceramic Membrane Reactors

In spite of the drawbacks of ceramic membranes for DH processes, namely their low separation selectivity to hydrogen, coke deposition, and their mechanical strength which is often insufficient for high temperature applications, a significant amount of experimental work has been reported on the subject of DH in ceramic membrane reactors [21, 27–29, 32, 40, 41] and there are some recent reports on ceramic membranes with high selectivity to H_2 (e.g., [26]).

A carbon membrane, made up of a bundle of 20 carbon fibers, was used in a reactor for the dehydrogenation of cyclohexane carried out at 195°C showing a fair improvement over equilibrium conversions [42]. A silicalite-supported zeolite membrane was used in a reactor for the dehydrogenation of isobutane achieving 50% isobutene yield increase due to equilibrium displacement. Similar yields were measured by sweeping the gas in co-current and countercurrent modes [28]. The conversion and selectivity of the metathesis of propene were altered by selectively removing trans-2-butene through a flat sheet stainless steel-supported silicalite-1 membrane [43]. In isobutane dehydrogenation over a chromia alumina catalyst conducted in a molecular-sieve carbon membrane [23] the conversion achieved in the counter-current flow operation method reached 85% at 500°C, probably a result of membrane transport and nitrogen dilution. In the vacuum mode, where no dilution occurs, the maximal conversion obtained was 40% at 500°C which is 10%

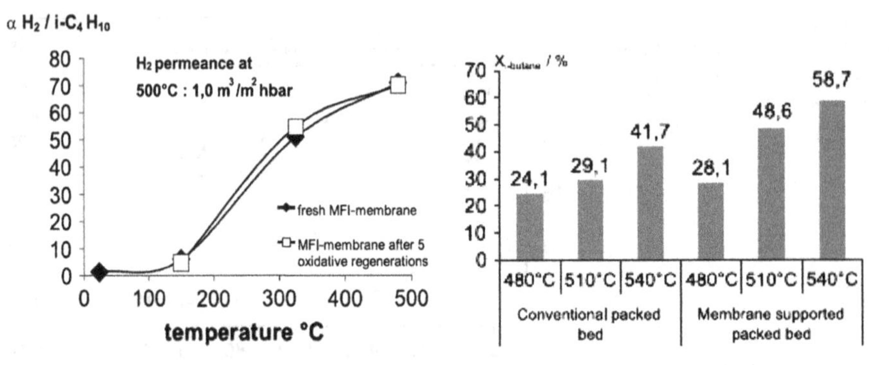

Fig. 9.8 *Left side* separation factor for a H_2/isobutane mixture on the MFI zeolite membrane as a function of temperature. *Right side* Isobutane conversion with and without removal of H_2 (after [29])

above the PFR conversion (Fig. 9.5). These results were compared with simulations that used the experimentally determined transport parameters.

Weyten et al. [32] reported implementation of a SiO_2 membrane prepared by CVD for propane DH. The membrane provided quite low hydrogen fluxes [$\sim$0.01 mol/(m^2 s) at 500°C] and a limited H_2/C_3H_8 permselectivity ($\alpha \approx$ 70–90 at 500°C). Yet, the propane conversion was improved by a factor of $\sim$2 above the equilibrium value ($\sim$18% at 500°C). Araki et al. [26] reported fabrication of SiO_2 membranes by CVD; the membranes provided similar hydrogen fluxes at much lower temperatures ($\sim$300°C) and exhibited very promising selectivities (24,000 H_2-to-CH_4 and 1,200 H_2-to-CO_2). Schäfer et al. [27] prepared SiO_2 membranes by a sol–gel process and tested them for propane DH over Cr_2O_3/Al_2O_3 and Pt–Sn/Al_2O_3 catalysts. Since low separation selectivity was obtained and the membranes suffered from coking, only minor improvement in propane conversion was achieved.

Zeolite membranes are another extensively studied ceramic membrane type [28, 29]. They are synthesized as a thin layer on a porous support (e.g., Al_2O_3). These membranes have H_2 permeation and separation selectivities similar to those of SiO_2 membranes, which again results in only minor improvement in DH conversions (Fig. 9.8). Also, there are experimental evidences of back-permeation of a sweep gas (N_2 [29]).

9.3.6 Membrane Reactor Modelling

Many studies have successfully simulated the observed behavior, in various membrane reactors and using independent kinetics and transport information; the applied models were too specific and complex to allow a general analysis. The models commonly assume plug flow and isothermal conditions on the tube and shell sides. Simulations included the dehydrogenation of 1-butene in a dense-membrane reactor [44], of n-butane in a porous-membrane reactor [45], or of isobutane in a Pd-membrane reactor ([23], Fig. 9.3), of ethane dehydrogenation in a PBR [46] or in a perspective membrane reactor [47]. Several groups have studied various design alternatives. Studies of radial dispersion effects in porous-membrane reactors, using the cyclohexane dehydrogenation kinetics, concluded that radial effects are important in high-permeability membranes [48]. The asymptotes and design of a packed tube and shell membrane reactor for the simple dehydrogenation reaction capitalizing on its fast reaction asymptote was analyzed in [38]. In that case, when local equilibrium is achieved for most of the reactor, the conversion shifts only due to hydrogen separation.

9.3.7 Thermal Management

Dehydrogenation is a highly endothermic process; therefore, heat has to be continuously supplied to drive the DH reaction. In industrial application, the heat is supplied by the heat exchange with other units, where the heat is generated by combustion; the heat realized during the regeneration step is also used. Experimental lab-scale studies on membrane DH reactors are normally performed under isothermal conditions, with the heat supplied by electrical heaters [e.g., 33, 37]. In theoretical studies, isothermal conditions are commonly applied [e.g., 28, 30] and the thermal effects were paid a little attention with regard to membrane DH reactors.

Several reports suggested oxidation of the separated hydrogen with a sweep gas (air or oxygen) catalyzed by Pd-based membrane to provide the required heat [30, 31, 38]. Compensation of DH endothermicity by oxidizing separated H_2 using air or oxygen as a sweep gas has been experimentally demonstrated by Gobina and Hughes [30] and Itoh and Wu [31]; Sheintuch [38] reported a model-based analysis of this approach. ODH is a proven industrial process, but there is only limited amount of literature on implementation of this approach in membrane reactors [39]. Though ODH allows effective heat supply, the DH selectivity is restricted. In membrane reactors, additional drawbacks may arise, such as membrane mechanical stability and durability.

9.4 Future Perspectives

The first decision to be made in designing a membrane reactor for a DH reaction is the membrane choice: the insufficient selectivities obtained with porous ceramic membranes and the high permeabilities obtained with Pd-composite membranes suggest the latter to be the best choice for a membrane. This indeed, may be an expensive solution, as discussed below, and the quest for other avenues should be pursued.

The second decision concerns the design of integrating the catalyst and the membrane. While catalytic reactor design is usually based on a rate expression developed in the laboratory, hydrogen separation by the membrane affects the rate, the selectivity, and the deactivation rate. Several works suggest that a minimal hydrogen partial pressure should be maintained to avoid these shortcomings, but there is no analytical tool to determine that value. This decision will affect other engineering choices like the mode of hydrogen removal from the membrane (by sweeping it or by collecting it undiluted at 1 bar or under vacuum) and the flow direction in the bed and in the membrane (co- or counter-current). Kinetic work aimed at deriving rate expressions at low hydrogen pressures should be conducted. The emergence of new computational catalysis tools may be helpful in that direction.

The third choice to be made is the mode of heat supply to sustain the endothermal DH reaction. ODH is one choice that at present yields unsatisfactory yields and selectivities. Other approaches are coupling the catalytic bed with another, in which an exothermic reaction is conducted. The authors have demonstrated such a solution, for hydrogen production in methane steam reforming, in a reactor composed of three concentric tubes (for methane combustion, for methane steam reforming, and for the Pd membrane) [49, 50]. Another approach is partial combustion of hydrogen on the membrane side. Very little work has been conducted on thermal solutions of membrane reactors.

Since the feasibility and durability of the hydrogen separation membranes and the feasibility of membrane reactor has been already demonstrated, the main obstacle for implementation is the cost added by the membrane. The cost contribution to the membrane can be gauged as follows: as stated previously (Sect. 9.3.2), the membrane area should satisfy a condition of $Pe_M = F_{A0}\Big/\left(J_{H_2}^M S^M\right) < 1$, where J_{H_2} is the membrane flux and S^M is its area, in order not to limit the rate. Taking $Pe_M = 0.5$ and using the best membrane permeabilities we can translate it to membrane surface area. The required catalyst amount is:

$$W_c = F_{A0}/r_{A0},$$

in order of magnitude, where r_{A0} is the initial rate. Thus, the ratio of membrane area to catalyst weight required is:

$$S^M/W_c = 2r_{A0}\Big/J_{H_2}^M$$

The commercial cost of Pd–Ag-supported membranes (mass production) is projected to be $\sim$1500 \$/m^2, and the catalyst cost for dehydrogenation applications is <100 \$/kg. The Pd–Ag membranes may provide fluxes (J_{H_2}) as high as 0.5–2.5 mol/(m^2 s) [16–18] and the DH reaction rates are in the range of $\sim$0.1–1 mol/(kg s). This estimation suggests that the membrane cost (when using mass production) will be comparable with the catalyst cost.

References

1. Weckhuysen BM, Schoonheydt RA (1999) Alkane dehydrogenation over supported chromium oxide catalysts. Catal Today 51:223–232
2. Cutrufello MG et al (2005) Preparation, characterization and activity of chromia-zirconia catalysts for propane dehydrogenation. Thermochim Acta 434:62–68
3. Cabrera F, Ardissone D, Gorriz OE (2008) Dehydrogenation of propane on chromia/alumina catalysts promoted by tin. Catal Today 133–135:800–804
4. Kogan SB, Herskowitz M (2001) Selective propane dehydrogenation to propylene on novel bimetallic catalysts. Catal Commun 2:179–185
5. Li W et al (2006) Propane dehydrogenation to propylene over Pt-based catalysts. Catal Lett 112:197–201
6. Lyu KL, Gaidi NA, Gudkov BS, Kostyukovskii MM, Kiperman SL, Podklenova NM, Kogan SB, Bursian NR (1986) Investigation of the kinetics and mechanism of the dehydrogenation of isobutane on platinum and platinum-indium catalysts. Kinet Katal 27:1371–1377
7. Lyu KL, Gaidi NA, Kiperman SL, Kogan SB (1990) Kinetics of dehydrogenation of n-butane on platinum catalysts. Kinet Katal 31:483–486
8. Murata K et al (2003) Dehydrogenation of propane over a silica-supported gallium oxide catalyst. Catal Lett 89:213–217
9. Murata K et al (2005) Dehydrogenation of propane over a silica-supported vanadium oxide catalyst. Catal Lett 102:201–205
10. Grabowski R (2006) Kinetics of oxidative dehydrogenation of C_2–C_3 alkanes on oxide catalysts. Catal Rev 48:199–268
11. Khodakov A (1999) Structure and catalytic properties of supported vanadium oxides: support effects on oxidative dehydrogenation reactions. J Catal 181:205–216
12. Cavani F, Ballarini N, Cericola A (2007) Oxidative dehydrogenation of ethane and propane: how far from commercial implementation? Catal Today 127:113–131
13. Blasko T (1997) Oxidative dehydrogenation of ethane and n-butane on VO_x/Al_2O_3 catalysts. J Catal 169:203–211
14. Buyanov RA, Pakhomov NA (2001) Catalysts and processes for paraffin and olefin dehydrogenation. Kinet Catal 42:72–85
15. Sheintuch M, Dessau RM (1996) Observation, modeling and optimization of yield, selectivity and activity during dehydrogenation of isobutene and propane in a Pd membrane reactor. Chem Eng Sci 51:535–547
16. Tong J et al (2006) Simultaneously depositing Pd-Ag membrane on asymmetric porous stainless steel tube and application to produce hydrogen from steam reforming of methane. Ind Eng Chem Res 5:648–655
17. Mori N et al (2007) Reactor configuration and concentration polarization in methane steam reforming by a membrane reactor with a highly hydrogen-permeable membrane. Ind Eng Chem Res 46:1952–1958
18. Chen Y et al (2008) Efficient production of hydrogen from natural gas steam reforming in palladium membrane reactor. Appl Catal B 80:283–294
19. Saracco G, Specchia V (1994) Catalytic inorganic membrane reactors: present experience and future opportunities. Catal Rev Sci Eng 36:305–384
20. Hsieh HP (1991) Inorganic membrane reactors. Catal Rev Sci Eng 33:1–70
21. Collins JP et al (1996) Catalytic dehydrogenation of propane in hydrogen permselective membrane reactors. Ind Eng Chem Res 35:4398–4405
22. Sznejer GA, Efremenko I, Sheintuch M (2004) Carbon membranes for high temperature gas separations: experiment and theory. AIChE J 50:596–610
23. Sznejer G, Sheintuch M (2004) Application of a carbon membrane reactor for dehydrogenation reactions. Chem Eng Sci 59:2013–2021
24. Zhang X et al (2006) Methanol steam reforming to hydrogen in a carbon membrane reactor system. Ind Eng Chem Res 45:7997–8001

25. Brunetti A et al (2004) WGS reaction in a membrane reactor using a porous stainless steel supported silica membrane. Chem Eng Process 46:119–126
26. Lee D et al (2004) Synthesis, characterization, and gas permeation properties of a hydrogen permeable silica membrane supported on porous alumina. J Membr Sci 231:117–126
27. Schäfer R et al (2003) Comparison of different catalysts in the membrane-supported dehydrogenation of propane. Catal Today 82:15–23
28. Casanave D et al (1999) Zeolite membrane reactor for isobutene dehydrogenation: experimental results and theoretical modeling. Chem Eng Sci 54:2807–2815
29. Illgen U et al (2001) Membrane supported catalytic dehydrogenation of iso-butane using an MFI zeolite membrane reactor. Catal Commun 2:339–345
30. Gobina E, Hughes R (1996) Reaction coupling in catalytic membrane reactors. Chem Eng Sci 51:3045–3050
31. Itoh N, Wu T-H (1997) An adiabatic type of palladium membrane reactor for coupling endothermic and exothermic reactions. J Membr Sci 124:213–222
32. Weyten H et al (2000) Membrane performance: the key issues for dehydrogenation reactions in a catalytic membrane reactor. Catal Today 56:3–11
33. Guo Y et al (2003) Preparation and characterization of Pd-Ag/ceramic composite membrane and application to enhancement of catalytic dehydrogenation of isobutene. Sep Purif Technol 32:271–279
34. Shu J, Granjean BPA, Van Neste A, Kaliaguine S (1991) Catalytic palladium based membrane reactors: a review. Can J Chem Eng 69:1036–1058
35. Hsieh HP (1996) Inorganic membranes for separation and reaction. Elsevier, Amsterdam
36. Gryaznov VM, Orekhova NV, Cybulski A, Moulijn JA (1998) Structured catalysts and reactors. Marcel Dekker, New York, pp 435–461
37. Liang W, Hughes R (2005) The catalytic dehydrogenation of isobutene to isobutene in a palladium/silver composite membrane reactor. Catal Today 104:238–243
38. Sheintuch M (1998) Design of membrane dehydrogenation reactors: the fast reaction asymptote. Ind Eng Chem Res 37:807–814
39. Raybold TM, Huff MC (2000) Oxidation of isobutene over supported noble metal catalysts in a palladium membrane reactor. Catal Today 56:35–44
40. Battersby S et al (2006) An analysis of the Peclet and Damkohler numbers for dehydrogenation reactions using molecular sieve silica (MSS) membrane reactors. Catal Today 116:12–17
41. Weyten H et al (1997) Dehydrogenation of propane using a packed-bed catalytic membrane reactor. AIChE J 43:1819–1827
42. Itoh N, Haraya K (2000) A carbon membrane reactor. Catal Today 56:103–111
43. Van de Graaf JM, Zwiep M, Kapteijn F, Moulijn JA (1999) Application of a zeolite membrane reactor in the metathesis of propene. Chem Eng Sci 54:1441–1445
44. Itoh N, Govind R (1989) Combined oxidation and dehydrogenation in a palladium membrane reactor. Ind Eng Chem Res 28:1554–1557
45. Gokhale YG, Noble RD, Falconer JL (1993) Analysis of a membrane enclosed catalytic reactor for butane dehydrogenation. J Membr Sci 77:197–206
46. Tsotsis TT, Chamagnie AM, Vasiteiadis SP, Ziaka ZD, Minet RG (1992) Packed bed catalytic membrane reactors. Chem Eng Sci 47:2903–2908
47. Gobina E, Hou K, Hughes R (1995) Equilibrium shift in alkane dehydrogenation using a high temperature catalytic membrane reactor. Catal Today 25:365–370
48. Koukou MK, Papayannakos N, Markatos NC (1996) Dispersion effects on membrane reactor performance. AIChE J 42:2607–2615
49. Simakov DSA, Sheintuch M (2009) Demonstration of a scaled-down autothermal membrane methane reformer for hydrogen generation. Int J Hydrogen Energy 34:8866–8876
50. Simakov DSA, Sheintuch M (2010) Experimental optimization of an autonomous scaled-down membrane reformer for hydrogen generation. Ind Eng Chem Res 49:1123–1129

Chapter 10
Steam Reforming of Natural Gas in a Reformer and Membrane Modules Test Plant: Plant Design Criteria and Operating Experience

Marcello De Falco, G. Iaquaniello and A. Salladini

10.1 Introduction

The natural gas steam reforming process is controlled by chemical equilibrium and significant hydrogen yields are achieved only at high temperatures (850–900°C). A part of methane feedstock has to be burned in furnaces, reducing the process global efficiency, increasing the greenhouse gas (GHG) emissions and strengthening the dependence of hydrogen cost on the natural gas cost.

Technology development in the past years has been focussed on reducing the reforming temperatures, in order to increase process efficiency and reduce operating costs. The integration of hydrogen selective membranes appears to be a promising way to enhance hydrogen yield at lower temperatures, because the selective removal of hydrogen produced in reaction zone prevents the equilibrium conversion to be achieved. Therefore, the continuous removal of the hydrogen from reaction environment should allow the reaction to be supported at lower temperature.

Among the typologies of hydrogen selective membranes, thin Pd-based supported membranes seem to be the most promising thanks to the high selectivity and high permeation flow [1–6]. They can be integrated in steam reforming process by means of two potential configurations:

M. De Falco (✉)
Faculty of Engineering, University Campus Bio-Medico of Rome, via Alvaro del Portillo 21, 00128 Rome, Italy
e-mail: m.defalco@unicampus.it

G. Iaquaniello
Tecnimont-KT S.p.A, Viale Castello della Magliana 75, 00148 Rome, Italy
e-mail: Iaquaniello.G@tecnimontkt.it

A. Salladini
Processi Innovativi S.r.l., Corso Federico II 36, 67100, L'Aquila, Italy
e-mail: salladini.a@processiinnovativi.it

M. De Falco et al. (eds.), *Membrane Reactors for Hydrogen Production Processes*,
DOI: 10.1007/978-0-85729-151-6_10,

1. Hydrogen selective membrane is assembled in separation modules applied downstream to reaction units. The process scheme, called reformer and membrane module (RMM), is composed by a series of reaction–separation modules [7, 8].
2. Selective membrane is assembled directly inside the reaction environment, so that the hydrogen produced by reactions is immediately removed. Such a configuration is called Membrane Reactor (MR) [9–13].

In a previous work, the authors have compared benefits and drawbacks of these configurations [14]. The main outcomes were:

- Pd-based membrane has to operate at temperature below 500°C for stability problem. This leads to limit reaction operating temperature in MR, while in RMM configuration operating temperatures in reactor and in separator are decoupled and can be fixed separately
- Globally, the hydrogen production costs for the RMM technology are more than 10% lower than those of the conventional H_2 scheme coupled with the cogeneration unit, where electrical power is produced by process outlet mixture.

Starting from these considerations, an innovative 20 Nm^3/h prototypal RMM plant has been designed and realized. The plant is composed by two-step reformers, working at 550–650°C, and two membrane modules at 450°C.

10.2 Process Scheme Description

The process scheme is illustrated in Fig. 10.1 and consists of two-step RMM.

Natural gas from battery limits or from cylinders at 20 barg is introduced through the pressure regulator and flow controller to the feed desulfurisation (DS) reactor, where sulfur compounds are removed. The desulfurised feed is then mixed with steam in ratio ranging from 3 to 4. Mixed feed is preheated in the convection section and enters the first reforming stage which works at a temperature of 550°C or higher. The reformed gas product is cooled down at 450°C and enters the first separation module. A retentate, recycled to the second reformer stage and a mixture of H_2 plus sweeping steam, are produced. The stream flowing out of the second reformer stage is cooled down from 650 to 450°C and routed to the second separation module. H_2 from both modules are mixed together and sent to final cooling and condensate separation. Retentate from the second stage is sent to the flare. The pressure of both shell and permeate sides are controlled using a back pressure regulators. Both membrane modules are protected using a pressure relief regulator installed on the income lines.

All the vent points are connected to main vent system and routed to the flare.

The reaction of heat in both reforming steps is provided by two independent hot gas generators in order to set the reforming temperatures as required by the tests.

Figure 10.2 shows a bird-eye view of the industrial test plant completely erected which occupies an area of 1,000 m^2.

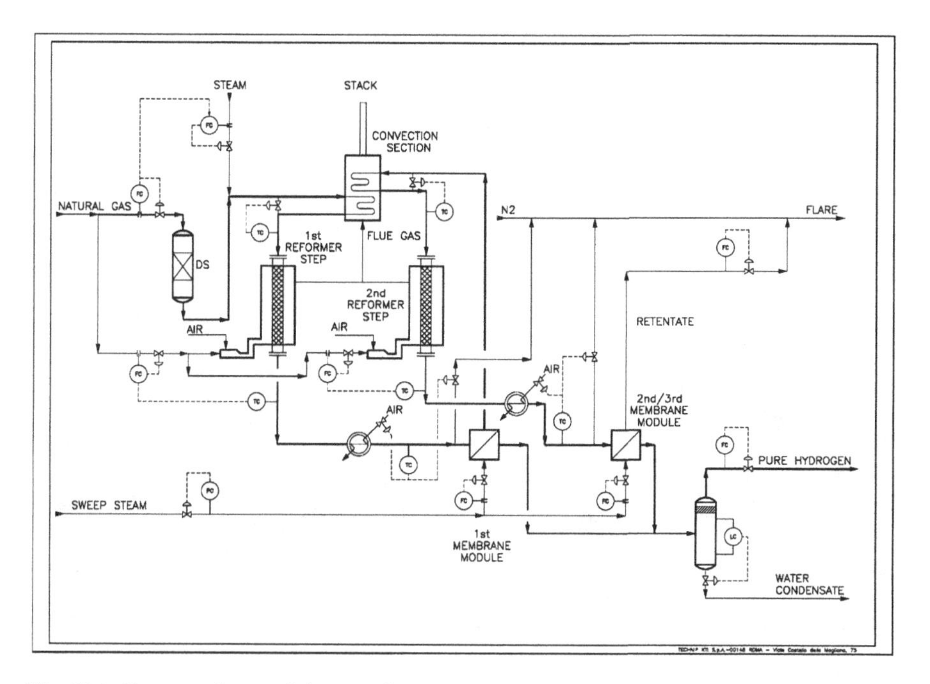

Fig. 10.1 Process scheme of the test plant

10.3 Reactors and Separators Design

10.3.1 Two-Steps Reformer Design

Each module of the reformer consists essentially of two main sections: the radiant box, containing the tube charged with the structured catalyst, and the convection section, where heat is recovered from the flue gases for preheating and superheating the feed and the steam. The steam is produced separately by a hot oil boiler. The steam reforming reactions are carried out at pressure up to 20 barg and temperatures in the range of 550–650°C, while the flue gases coming from the hot generator may reach a temperature >800°C. Design of the radiant chamber is quite conventional, it differs only by the heated length of the reformer tube which is around 3 m, the tube metallurgy and the contained catalyst. One of the advantages of the proposed architecture is to require low cost stainless steel instead of exotic and quite expensive material as HP25/35 chromium/nickel alloy.

Traditionally, in the steam reforming of natural gas there is a strong endothermic reaction region located at the inlet of the catalytic tubes. At such a section, the rate limiting process is the heat transfer from the combustion gases through the heated reactor wall and catalyst bed.

The lower reforming temperature rises the issue of how much CH_4 conversion is close to the equilibrium value. Preliminary experimental results [15] caused

Fig. 10.2 View of the industrial test plant

Fig. 10.3 Steam reforming catalyst

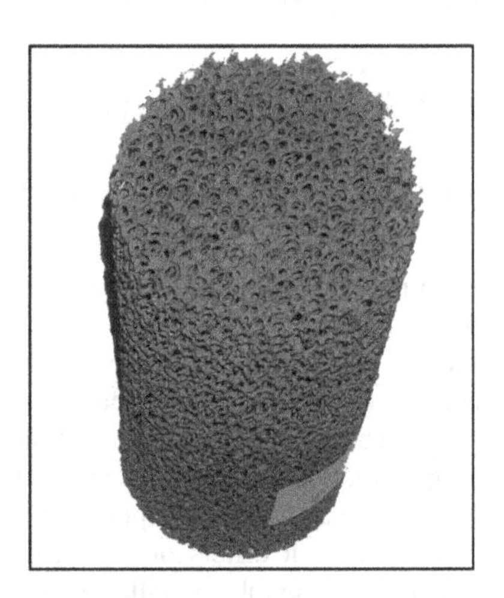

some doubts about the effectiveness of industrial nickel based catalyst supported on ceramic materials as α-Al_2O_3, MgO and Mg–Al_2O_4. In order to enhance the conductive heat transfer and the catalyst activity, a noble metal catalyst supported on SiC foam (see Fig. 10.3) has been investigated. The open cross flow SiC foam is expected to increase the effectiveness of convective heat transfer allowing the tube metal temperature (T_{mt}) to be reduced with the same heat flux or conversely allowing the heat flux to be increased with the same T_{mt}.

Table 10.1 Main properties and geometry of the membranes

Developer	Substrate Support	Membrane selective layer	Selective layer thickness (μm)	Membrane surface (m^2)	Manufacturing method	Geometry
ECN	Alumina	Pd	3–9	0.4	Electroless deposition	Tubular
Confidential	Alumina	Pd/Ag	2–3	–	Electroless deposition	Tubular
MRT	Stainless steel	Pd/Ag	25	0.6	Proprietary	Planar
ACKTAR	Stainless steel + alumina	Pd or Pd/Ag	3–5	0.13	Reactive sputter deposition	Tubular planar

The use of a foam as catalytic support results also in lower pressure drops in the catalytic bed which in turn reduce the energy costs associated with the gas compression and with the recirculation of gas streams as in the proposed architecture.

10.3.2 Dense Pd-based Membranes for Hydrogen Separation

In this section, a description of the four different membrane separators installed in the plant is given. Such units enable hydrogen separation at high temperature.

During the last decade a significant effort was devoted to thin or low thickness (<10 μm) Pd-based membranes supported on different porous substrates and to move the technology from lab scale to larger surface dimensions. Among the few membranes provider/developers, three of them, ECN from The Netherland [16], a company from Japan and MRT from Canada [17] were selected and included in the project. A fourth company, Acktar from Israel [18] was also involved, but its membrane will be tested later. All these membranes are Pd based, but with different supports. ECN membranes and those ones from Japanese manufacturer are made by a ceramic support whereas Acktar and MRT membranes have a stainless steel substrate. With the exception of ECN, all membranes are Pd/Ag alloy membranes. The membranes have a tubular geometry, with the exception of MRT which is planar.

Table 10.1 reports the main properties and geometry of membranes included in the project. Figure 10.4a–d shows the different modules.

10.4 Control System and Testing Strategies

The main features of the instruments of the control system are reported here. Flow controllers are used for natural gas, steam, sweeping steam to the membrane separator and hydrogen produced. Temperatures and pressures are measured and

Fig. 10.4 **a** ECN membrane separator (total surface = 0.4 m^2). **b** Three tubular membrane separator. **c** MRT flat membrane. **d** Acktar novel stainless steel alumina membrane support

recorded before and after each reformer and separation step. A differential pressure sensor monitors the pressure drop across the catalytic tube. CH_4, CO, and CO_2 concentrations are monitored by online infra-red NDIR multiple analyzer, ABB URAS 14, at the outlet of the two reforming steps and of the membrane modules outlets. H_2 is analyzed by a thermal conductivity detector ABB Caldos 17 and through a Perkin Elmer GC Autosystem XL. By acting on the reformer outlet temperature and on the steam/carbon ratio it is possible to generate different syngas compositions to be fed to the separation module and to analyze the role played by partial pressures of the main components (CH_4, CO, CO_2, and H_2) in the separation performance (see Fig. 10.5).

An overview of the syngas composition range used as feed in membrane testing is reported in Table 10.2; such compositions resulted from the natural gas conversion obtained in the two reforming steps. Alternatively feed gas composition could also be obtained by mixing up the pure components directly from the cylinders.

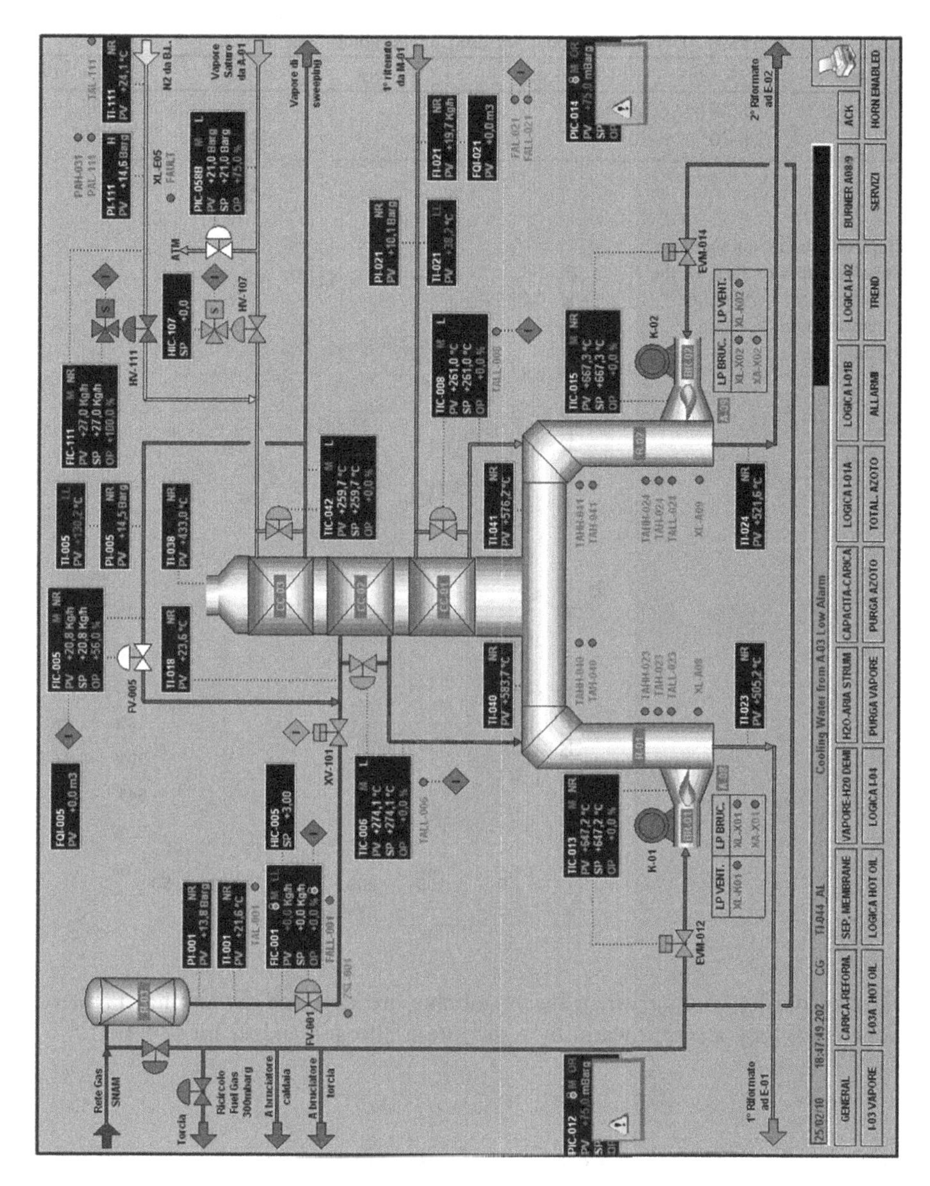

Fig. 10.5 Control philosophy of the pilot unit

It is also possible to modify the inlet temperature to the separation modules in the range 400–500°C by acting on the cooling media flow of the pre-heater installed just before the module. On the permeate side, steam is added as sweeping media to lower the partial pressure of hydrogen and to increase the driving force for separation.

Heating and cooling of the membrane are carried out under nitrogen atmosphere, as suggested by the producers. In order to reduce the flushing time with

Table 10.2 Overview of syngas composition used in membrane testing

Mixture	H_2% vol	CH_4% vol	CO_2% vol	CO%vol	H_2O% vol
1	28–30	6–8	6–9	1–2	54–55
2	24–26	7–9	6–8	1–2	58–59

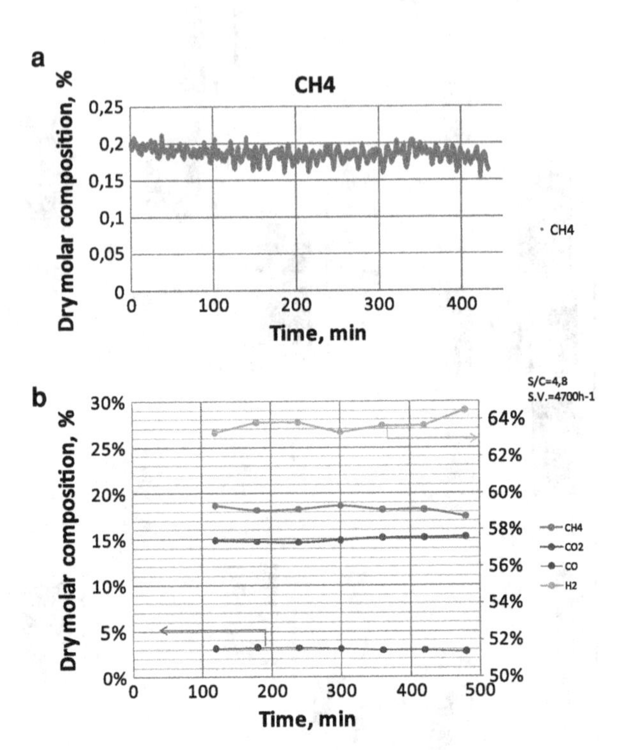

Fig. 10.6 **a** Instantaneous measurements of CH_4 in the reformed gas. **b** Average measurements of main components

nitrogen during cooling, when the membrane are not used for testing (night or week-end), they are kept warm by a mixture a nitrogen and steam at 400°C.

10.5 Process Conditions for Plant Testing

A wide variety of operating conditions is adopted in the tests performed to assess the proper operability in an industrial environment and to have a clear evidence of the potential of RMM technology in terms of efficiency. Figure 10.6a and b report gas composition monitored on-line; in particular Fig. 10.6a shows instantaneous measurements of the methane at the outlet of the first reforming step. Figure 10.6b represents the hourly average for the main components at a pressure of 10 barg and a temperature of 600°C at the outlet of the first reforming step.

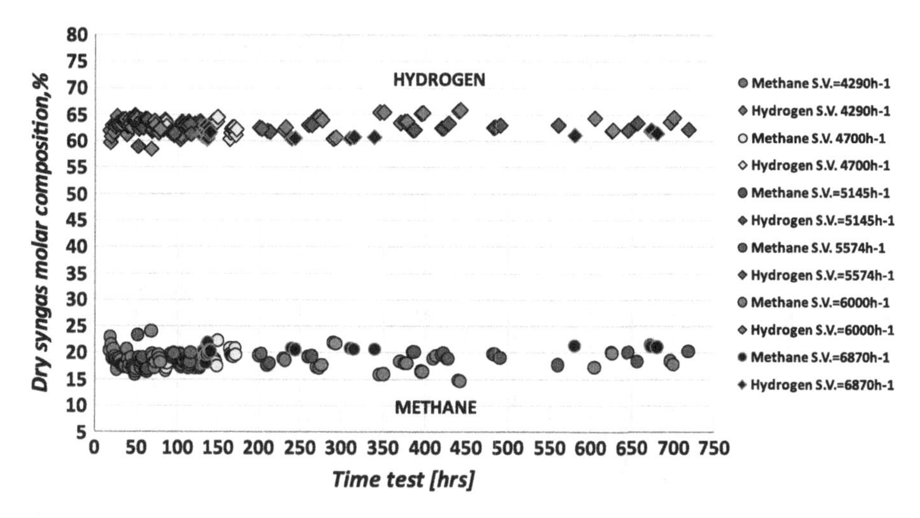

Fig. 10.7 Methane and hydrogen content in the dry product along 720 h operation

10.5.1 Low Temperature Steam Reforming Testing

In order to obtain a full set of data, several tests were carried out by varying the reformer outlet temperature from 550 to 650°C and pressure from 10 to 20 barg. The experimental points in Fig. 10.7 represent methane and hydrogen contents in the syngas during a steady state operation at 550°C, S/C of 4.8 and 10 barg during 720 h on stream. Although the on-stream time is limited no sign of deactivation has been yet detected.

Methane conversion versus the gaseous inlet hourly space velocity is shown in Fig. 10.8. Conversion efficiency is definitely affected by the space velocity (s.v.). Converted methane increases with s.v. as result of a better conductive heat and mass transfer up to a peak at around 5,000 h^{-1}. Higher s.v. reduce the methane conversion from 49.5 to 46.5%.

10.5.2 Temperature Variations on Membrane Testing

It is assumed that Sieverts' law is verified and consequently:

$$J_{H_2} = \frac{\text{Perm}}{\delta} \cdot \left(p_{H_2,\text{up}}^{0.5} - p_{H_2,\text{down}}^{0.5}\right) \tag{10.1}$$

where J_{H_2} is the hydrogen flux permeated, δ is the membrane thickness, $p_{H_2,\text{ up}}$ and $p_{H_2,\text{ down}}$ are hydrogen partial pressures up and downstream. Then, inlet temperature at the membrane separation modules is changed within the range 400–500°C in order to define membrane permeability expression.

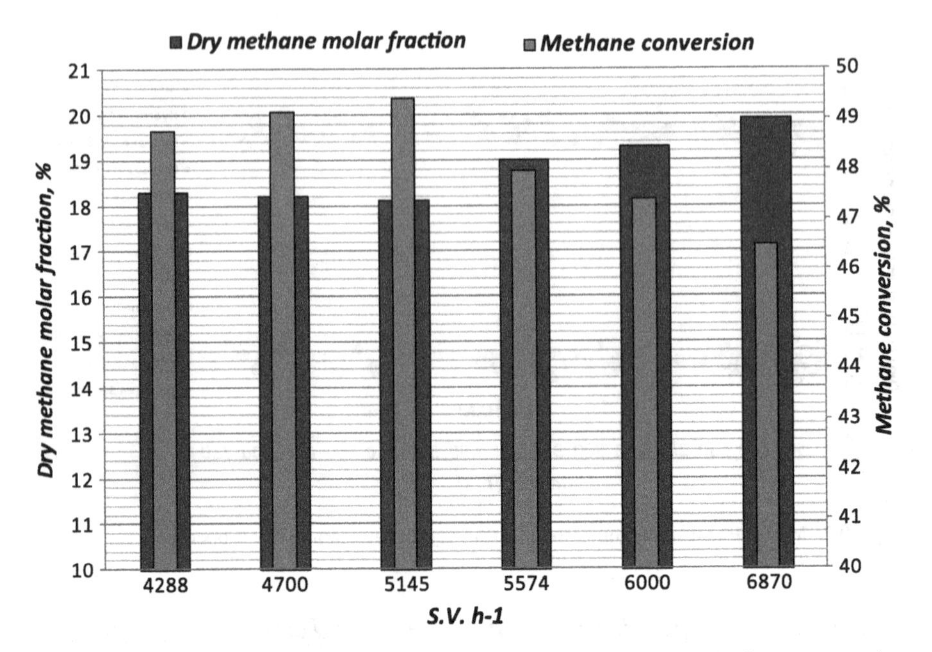

Fig. 10.8 Methane conversion versus gaseous (inlet) hourly space velocity at an outlet temperature of 600°C, at a pressure of 10 barg and S/C of 4.8

It is well-known that hydrogen flux through the membrane usually increases with the temperature, since membrane permeability depends on operating temperature according to Arrhenius' law (refer to Chap. 2):

$$\text{Perm} = A \cdot \exp\left(\frac{-E_\text{a}}{\text{RT}}\right) \tag{10.2}$$

Values of parameters A and $E_a$for the membranes assembled can be found by plant testing through the following procedure:

1. At a fixed pressure difference between shell and lumen (10 barg in retentate, 0.4 barg in permeate zone), the hydrogen flux is measured for different operating temperatures.
2. Assuming the Arrhenius permeability dependence on the temperature and Sieverts' law, the hydrogen flux can be described by the following expression:

$$J_{\text{H}_2} = \frac{A \cdot \exp\left(\frac{-E_\text{a}}{\text{RT}}\right)}{\delta} \cdot \left(p_{\text{H}_2,\text{up}}^{0.5} - p_{\text{H}_2,\text{down}}^{0.5}\right) \tag{10.3}$$

Equation (10.3) can also be written as

$$\frac{J_{\text{H}_2}}{\left(p_{\text{H}_2,\text{up}}^{0.5} - p_{\text{H}_2,\text{down}}^{0.5}\right)} = \text{Permeance} = \frac{A \cdot \exp\left(\frac{-E_\text{a}}{\text{RT}}\right)}{\delta} \tag{10.4}$$

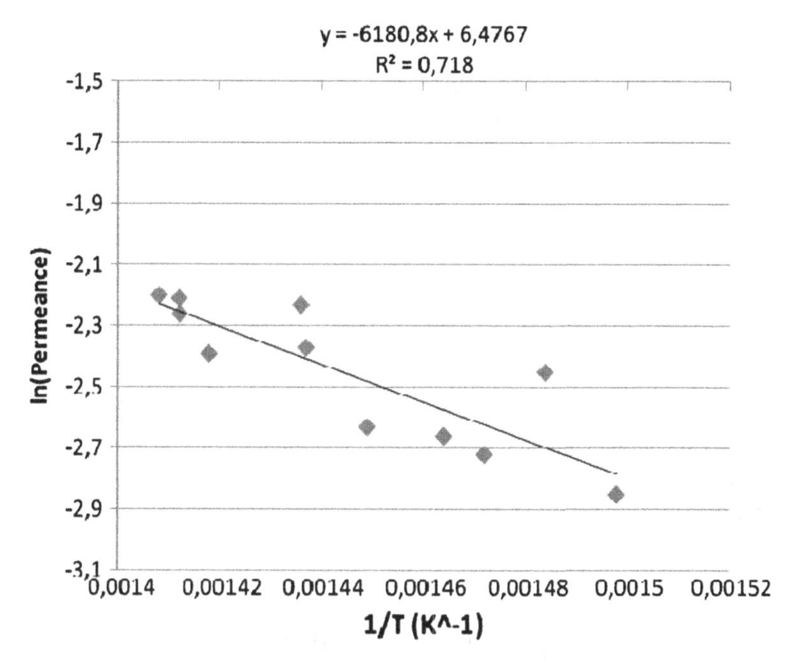

Fig. 10.9 ln (permeance) versus 1/T for a Pd-based membrane

Equation (10.4) is linearized as follows:

$$\ln(\text{Permeance}) = C_1 + \frac{C_2}{T} \quad (10.5)$$

where

$$C_1 = \ln\left(\frac{A}{\delta}\right) \qquad C_2 = -\frac{E_\mathrm{a}}{R} \quad (10.6)$$

3. The data collected are reported as $\ln(\text{Permeance})$ versus $1/T$. If the data arranged on a straight line, Arrhenius dependence on the temperature is verified and the values of A and E_a can be derived from the slope and the intercept of the fitting straight line.

Figure 10.9 reports experimental data for one of the membranes tested (2.5 μm thick) and the linear regression expression. A good correlation is obtained ($R^2 = 0.718$).

Thus the membrane permeability expression is calculated:

$$\text{Perm} = 1.62 \cdot 10^{-3} \cdot \exp\left(\frac{-51362}{\text{RT}}\right) \frac{\text{kmol}}{\text{m} \cdot \text{h} \cdot \text{kPa}^{0.5}} \quad (10.7)$$

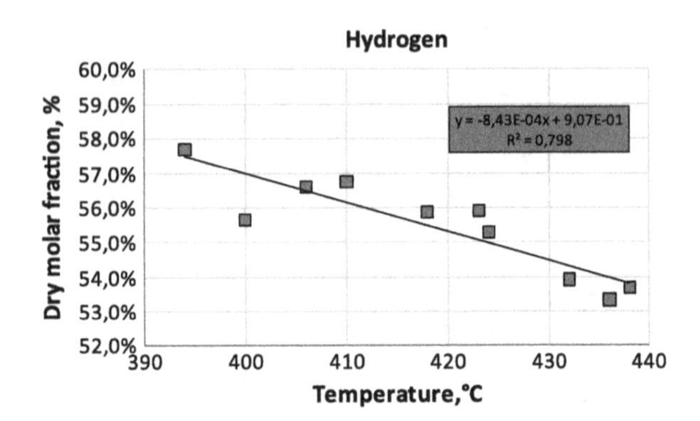

Fig. 10.10 Hydrogen content in the retentate side of the separation module versus the operating temperature

Figure 10.10 shows the variation of the hydrogen flux versus the operation temperature of the membrane module. By raising the temperature from 400 to 440°C the hydrogen content in the retentate is reduced from 57% to < 54% vol.

10.5.3 Feed and Sweep Flows on Membrane Testing

Feed flow impact on membrane performance is analyzed in the range 25–110% of the design flow. It is expected that an increase of feed flow will reduce the feed conversion. In Fig. 10.11 methane conversions with and without membrane are compared versus the space velocity after 70 h of operation with ECN membrane without steam sweeping. It appears quite clearly that by increasing the space velocity conversion of methane decreases.

10.6 Plant Global Performance

Global performance of the new architecture is presented in Fig. 10.12 in terms of methane conversion.

By comparing the methane content at the outlet of the first reaction step, R-01, with the content at R-02, is possible to quantify the methane conversion increase from 47.3% at R-01 outlet to 60% at R-02 outlet: a 25% increase at the same outlet temperature and S/C.

Effect of membrane on the methane concentration in the retentate gas can be analyzed looking at the difference between the outlet of R-01, 19% average, and 23% at the outlet of first separation module.

Such effect appears to be less important (from 14 to 15.5%) in the second separation module, mainly due to the fact that the second module surface was 1/5 of the first.

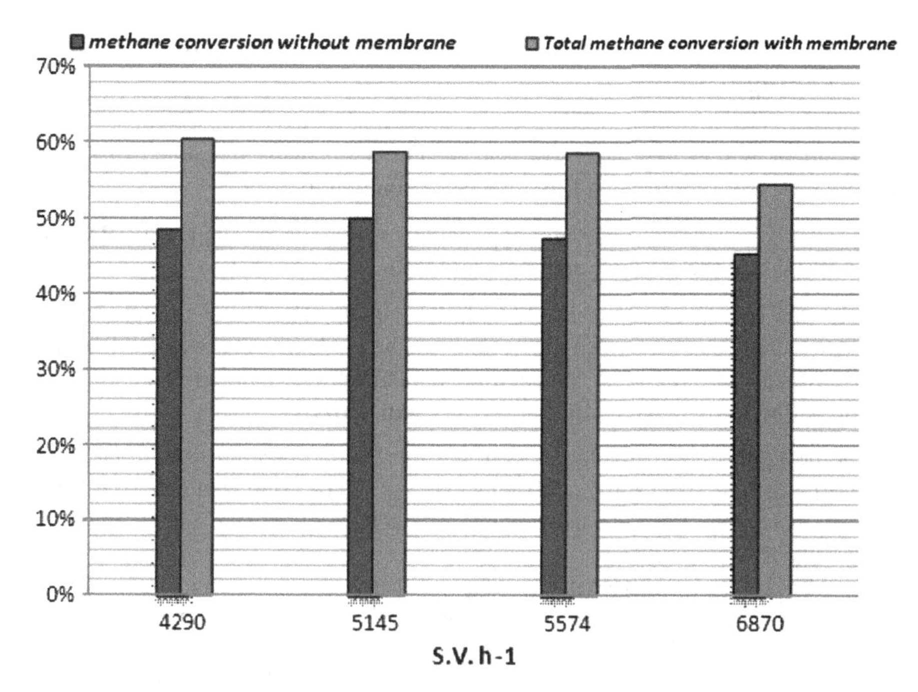

Fig. 10.11 Comparison of methane conversion with and without membrane versus the space velocity

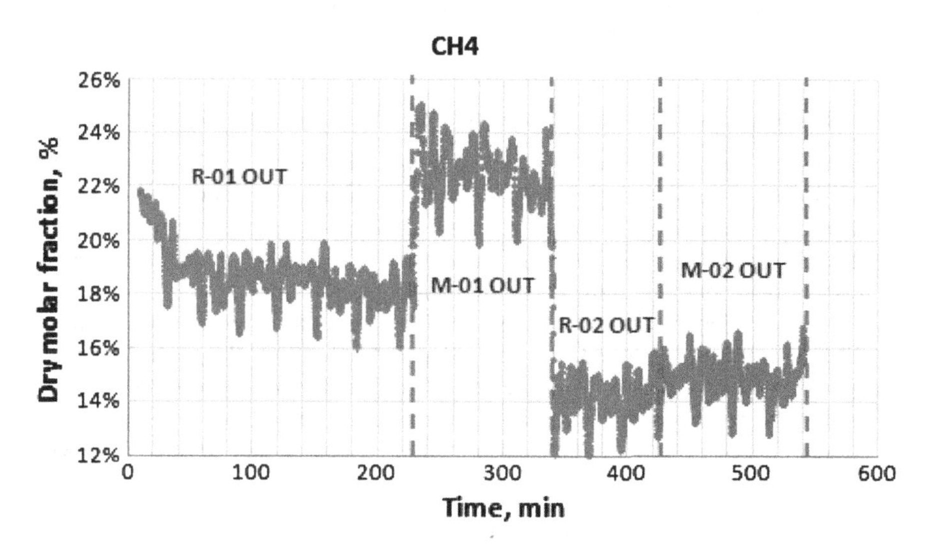

Fig. 10.12 CH_4 content for integrated process (reactors pressure = 10 barg, temperature = 600°C, S/C = 4.8): Reaction (R-01), H_2 Separation (M-01), Reaction (R-02), H_2 Separation (M-02)

10.7 Comparison of RMM and Integrated Membrane Reactor Configurations

As reported before, hydrogen selective membranes can be integrated in steam reforming process by means of two potential configurations:

- assembled in separation modules applied downstream to reaction units (RMM);
- directly inside the reaction environment (MR).

After the testing phase on RMM prototypal plant reported in this chapter, a comparison between the two configurations can be made in order to assess benefits and drawbacks in integrating selective membrane internally or externally.

The RMM benefits over MR can be thus summarized:

1. The possibility to decouple reaction and separation operating conditions. The reforming and separation module temperatures can be optimized independently, both increasing methane conversion for each reaction step and membrane stability and lifetime.
2. Simplification of the mechanical design of membrane tubes compared to those embedded in catalyst tubes. A simple "shell and tube" geometry can be selected for the tubular separation module.
3. Simplification of membrane modules maintenance and of catalyst replacement.

On the other hand, the main drawbacks are:

1. No compactness of the process, since RMM configuration is composed by reactors, separation modules, heat exchangers while, in MR, reaction and H_2 separation are performed in one single and compact device.
2. A higher membrane surface required.

At the same time, the main benefits of MR over RMM configuration are:

1. Compactness of the process. The integration in a single device of steam reforming reaction and hydrogen separation step leads to a more compact reforming plant.
2. Easiness in scalability. Scale-up and scale-down of the membrane reformer are very easy through an increase or a decrease of a number of parallel tubular reactors or of the length of the single membrane reformer.
3. Less application of useless catalyst. In the traditional process, due to strong radial temperature gradient, the catalyst pellets placed in the central zone of reformers usually do not work well since the temperature is too low for promoting the reactions. In membrane reformers, the central zone of the reactor does not contain catalyst but it is taken by the membrane tube devoted to collect the hydrogen permeated.

Main drawbacks of MR configuration are:

1. Technological problems in designing the reactor. A sensible component as the selective membrane has to be inserted in a critical environment.

2. The coupling between reaction and separation operating conditions, leading to the necessity to find a compromise between the reaction and the separation requirements.
3. No easy maintenance of the reactor.

At the current selective membranes state-of-the-art, the advantage of separately fixing reactors and separators operating conditions seems to be crucial, considering the sensitivity of supported Pd-based membranes to temperature. Therefore, RMM plant can be considered the most reliable solution.

A further improvement of membrane technology, foreseeable in the next years, should change this perspective, promoting the integration of membranes directly inside the reactors.

10.8 Conclusions

The primary objective of this project was to provide clear evidence of potentialities of the selective membrane technology in major chemical, petrochemical and other process applications to provide substantial reductions in energy consumption and in investment costs. It was recognized that the construction of a semi-industrial test plant is required to assess the operability in an industrial environment. Therefore a 20 Nm^3/h hydrogen membrane steam reforming plant has been designed and fabricated and a complete control system has been integrated in order to monitor reactors and membranes performance varying operating conditions (temperatures, pressures, flow-rates, and sweeping flow).

In this chapter the design criteria implemented have been presented, together with the operating conditions ranges analyzed. A catalytic foam has been inserted in the reformers in order to increase the effectiveness of the heat transfer and to reduce pressure drop, and four types of Pd-based hydrogen selective membranes are tested and compared.

Preliminarily experimental results have been presented, confirming the interest in such a new architecture.

Acknowledgment This work was carried out within the framework of the project "Pure hydrogen from natural gas reforming up to total conversion obtained by integrating chemical reaction and membrane separation," financially supported by MIUR (FISR DM 17/12/2002)-Italy.

References

1. Dittmeyer R, Höllein V, Daub K (2001) Membrane reactors for hydrogenation and dehydrogenation processes based on supported palladium. J Mol Catal A Chem 173:135–184
2. Howard BH, Killmeyer RP, Rothenberger KS, Cugioni AV, Morreale BD, Enick RM, Bustamante F (2004) Hydrogen permeance of palladium-copper alloy membranes over a wide range of temperatures and pressures. J Membr Sci 241:207–218

3. Peachey NM, Snow RC, Dye RC (1996) Composite Pd/Ta metal membranes for hydrogen separation. J Membr Sci 111:123–133
4. Ozaki T, Zhang Y, Komaki M, Nishimura C (2003) Preparation of palladium-coated V and V-15Ni membranes for hydrogen purification by electroless plating technique. Int J Hydrogen Energy 28:297–302
5. Tosti S, Bettinali L, Castelli S, Sarto F, Scaglione S, Violante V (2002) Sputtered, electroless and rolled palladium-ceramic membranes. J Membr Sci 196:241–249
6. Tosti S, Basile A, Bettinali L, Borgognoni F, Chiaravalloti F, Gallucci F (2006) Long-term tests of Pd–Ag thin wall permeator tube. J Membr Sci 284:393–397
7. De Falco M, Barba D, Cosenza S, Iaquaniello G, Farace A, Giacobbe FG (2009) Reformer and membrane modules plant to optimize natural gas conversion to hydrogen. Special Issue of Asia-Pacific J Chem Eng Membr React DOI:10.1002/apj.241
8. De Falco M, Barba D, Cosenza S, Iaquaniello G, Marrelli L (2008) Reformer and membrane modules plant powered by a nuclear reactor or by a solar heated molten salts: assessment of the design variables and production cost evaluation. Int J Hydrogen Energy 33:5326–5334
9. Shu J, Grandjean B, Kaliaguine S (1994) Methane steam reforming in asymmetric Pd and Pd-Ag porous SS membrane reactors. Appl Catal A: Gen 119:305–325
10. De Falco M, Nardella P, Marrelli L, Di Paola L, Basile A, Gallucci F (2008) The effect of heat flux profile and of other geometric and operating variables in designing industrial membrane steam reformers. Chem Eng J 138:442–451
11. De Falco M, Di Paola L, Marrelli L, Nardella P (2007) Simulation of large-scale membrane reformers by a two-dimensional model. Chem Eng J 128:115–125
12. Lin Y, Liu S, Chuang C, Chu Y (2003) Effect of incipient removal of hydrogen through palladium membrane on the conversion of methane steam reforming: experimental and modelling. Catal Today 82:127–139
13. Chai M, Machida M, Eguchi K, Arai H (1994) Promotion of hydrogen permeation on a metal-dispersed alumina membrane and its application to a membrane reactor for steam reforming. Appl Catal A: Gen 110:239–250
14. De Falco M, Iaquaniello G, Cucchiella B, Marrelli L (2009) Syngas: production methods, post treatment and economics. Reformer and membrane modules plant to optimize natural gas conversion to hydrogen. Nova Science Publishers, Newyork. ISBN: 978-1-60741-841-2
15. Progetto FISR Vettore idrogeno-2002. Rapporto sulle strutture del processo innovativo, issued by Technip-KTI February 2006, available on http://www.fisrproject.com
16. http://www.hysep.com, Hysep 1308 module
17. http://www.membranereactor.com
18. http://www.acktar.com

Chapter 11
Future Perspectives

Marcello De Falco, Gaetano Iaquaniello and Luigi Marrelli

11.1 Current Technology Status

Selective membrane application in chemical processes represents one of the most interesting scientific and technological topics over the last years, in the context of industrial process intensification tendency aimed to improve process efficiency.

Many efforts devoted to the development of membrane competitive applications by the most prestigious research centers worldwide attest the strategic importance and the potentiality of membrane reactors for the industry. The scientific production dealing with selective membrane reactors is growing exponentially: as reported in Chap. 2, 750 papers on membrane reactors have been published in 2009, of which 220 on Pd-based membranes. The main processes in which R&D departments are focusing the attention are those devoted to hydrogen production, for which two configurations are under study:

1. *Integrated membrane reactor* configuration. Selective membrane is assembled directly inside the reaction environment, so that the hydrogen produced by reactions is immediately removed.
2. *Staged membrane reactor* configuration. Selective membrane is assembled in separation modules applied downstream to reaction units, and the plant is composed by a series of reaction-separation modules.

M. De Falco (✉) and L. Marrelli
Faculty of Engineering, University Campus Bio-Medico of Rome,
via Alvaro del Portillo 21, 00128 Rome, Italy
e-mail: m.defalco@unicampus.it

G. Iaquaniello
Tecnimont-KT, Viale Castello della Magliana 75, 00148 Rome, Italy
e-mail: Iaquaniello.G@tecnimontkt.it

M. De Falco et al. (eds.), *Membrane Reactors for Hydrogen Production Processes*,
DOI: 10.1007/978-0-85729-151-6_11,

All these efforts and resources are producing results so that more and more effective and suitable selective membranes are made and applied, even if some problems have to be still overcome before a wide industrial use.

Almost all fulfillments developed up today concern experimental apparatuses with very small productivities, and no scale-up strategies are conceived yet, mainly because only few companies are able to supply high-surface Pd-based membrane modules.

In Chap. 3, a review of the few manufacturers able to produce selective Pd-based membranes with surfaces suitable for pilot plants (0.1–0.5 m^2) has been reported. Five providers have been found:

1. The Energy Research Centre of the Netherlands (ECN) that produces membranes of Pd on alumina support with a surface area equal to 0.4 m^2 and selective layer thickness of 3–9 μm.
2. MRT, a Vancouver-based private company, produces membranes either as rolled foils or as deposited thin films (8–15 μm). Maximum surface is 0.6 m^2.
3. An important Japanese Company (JC) is developing a gas separation membrane of Pd on Al_2O_3 support. The surface of selective layer is 28.3 cm^2.
4. SINTEF, research and educational centre in the field of environmental technology, has developed a very thin Pd-alloy selective layer (approximately 2–3 μm) supported on macroporous substrates, able to operate at high pressure.
5. The Israel company ACKTAR is developing a 3–5 μm thin Pd or Pd–Ag membrane on steel substrate (refer to Chap. 10).

However, laboratory scale reactor performance and assessments by mathematical models simulations have shown the real and excellent potentiality of membrane integration in chemical processes, leading to a strong increase of reactant conversion at lower operating temperatures. Table 11.1 summarizes the main outcomes reported in Chaps. 5, 6, 7, 8, and 9), where some interesting case studies have been presented and described.

Obviously, membrane reactor performance is strongly dependent on selective membrane behavior in terms of permeability, selectivity, and stability. Chapter 2 reports several data about hydrogen permeation performance of different metal-based and particularly of Pd-based membranes.

Table 11.1 Membrane reactor performance for various chemical processes

Process	Membrane reactor configuration	Improvement in respect to traditional process	Reference
Natural gas steam reforming	Integrated membrane reactor	20–50%	Chap. 5
Autothermal reforming	Integrated membrane reactor	5–20%	Chap. 6
Water–gas shift	Integrated membrane reactor	10%	Chap. 7
Water–gas shift	Staged membrane reactor	7%	Chap. 7
H_2S cracking	Staged membrane reactor	10–200%	Chap. 8
Alkanes dehydrogenation	Integrated membrane reactor	50–300%	Chap. 9

Recently, the first membrane reactor pilot plant has been realized. A staged membrane reactor for natural gas steam reforming, also called reformer and membrane modules (RMM) test plant, having the capacity of 20 Nm^3/h of hydrogen, has been designed and constructed to investigate at an industrial scale level the performance of such innovative architecture.

The process operates at low thermal level (below 650°C in comparison to 850–950°C needed in tradition plants), whereas membrane modules work at 450°C, a safe temperature for Pd-based membrane stability. Data concerning plant behavior after 720 h are extensively reported in Chap. 10: it has been demonstrated that a final methane conversion after a two-step reaction equal to 60% can be obtained, equivalent to an improvement of 20% over the equilibrium threshold (conversion equal to about 50% at 650°C, 10 barg and a ratio between steam and methane feedstocks of 4.8), moreover recovering a highly pure hydrogen stream.

11.2 Barriers to Be Overcome

Although all these efforts have produced very promising results, some technological challenges still need to be addressed before hydrogen selective membranes, and particularly Pd-based membranes, will become enough reliable and cost-competitive for industrial applications. Of course, an interdisciplinary approach, with a stringent collaboration between industry and university, and among various scientific and technological sectors such as chemistry, physics, and engineering, is crucial to make the commercialization of hydrogen selective membranes and of membrane reactors a reality.

In the following sections, the main barriers to be overcome are listed and commented.

11.2.1 Fabrication Methods

The development of manufacture methods for producing thin selective layer membranes with high selectivity, high permeability, and high stability is the crucial task to be reached.

Main production methods are described in Chaps. 2 and 3, but no one is still completely developed in terms of costs and reliability and able to produce high surface (some square meters) membranes.

Concerning hydrogen selective membranes, industry is mainly focused on the manufacture of composite membranes made of a thin Pd film on porous substrates. Reducing the selective layer thickness allows membrane cost to be decreased (decreasing the Pd thickness by a factor two reduces the total Pd cost by a factor four) and increasing the hydrogen flux, which is in inverse proportion with the film thickness. On the other side, the porous substrates provides the mechanical strength needed for the application in chemical processes and for tolerating the down-up pressure difference, which is the hydrogen permeation mechanism driving force.

Two main challenges have to be tackled:

1. *The selection of porous substrate*. Ceramic supports are chemically stable, have small pore sizes and a more uniform pore size distribution, essential for the formation of thinner and uniform membrane layer. But they are brittle, and the large difference in thermal expansion coefficients between palladium layer and ceramic supports leads to membrane cracking and loss of adherence. On the other hand, metallic porous supports made of stainless steel (PSS) have a more similar thermal expansion coefficient, reducing thermal stresses, but the large pore size and wide pore size distribution make the formation of a uniform selective layer problematic.
2. *Selective layer deposition methods*. Depositing the Pd-based layer on selected substrates is not an easy task. Finding and applying reliable and cost-competitive deposition methods is the main challenge that scientists and technicians have to still overcome. Deposition method has to assure a proper membrane stability within the operative range in which the membrane has to work.

11.2.2 Membrane Surface Poisoning and Membrane Durability

Another critical problem is represented by the palladium surface contamination of Hg vapor, hydrogen sulfide, SO_2, thiophene, arsenic, unsaturated hydrocarbons, chlorine, and carbon from organic materials (refer to Chap. 2).

The lack of long-term durability data creates an uncertainty about membrane reliability for industrial application. Most long-term stability studies reported in the literature are in the order of hundreds of hours, whereas for an industrial application, long-term stability tests have to be carried out in thousands of hours [1].

Taking as a reference the targets imposed by the US DOE, the stability/durability target for 2007 is 3 years, 2010 target is 7 years, and 2015 target is $>$10 years [2]. This means that a long road has to be still covered.

11.2.3 Membrane Integration in Reaction Environment

Integrating the selective membrane in a critical environment as chemical reaction one (high temperature and pressure, presence of membrane contaminants), i.e., imposing the integrated membrane reactor configuration, is a challenging issue. Nowadays, only laboratory scale prototypes have been constructed, and even if promising performance has been obtained, the reactor scale-up problems have still not be properly faced.

Overcoming these barriers should lead to make Pd-based hydrogen selective membrane an industrial product in a few years.

11.3 Technological Perspectives

This section provides some reflections and focus on future work and data required to advance the membrane technology toward commercial application. Best solutions to overcome barriers listed in the previous paragraph (membrane architecture and manufacturing, interaction between the membranes and major components of the syn-gas, long-term durability) are here proposed.

11.3.1 Membrane Architecture and Manufacturing

High permeation flux is the key determinant for future commercial application. As aforementioned, thin, few μm thickness, Pd or Pd alloy–based films supported on porous substrates are the right answer, providing the needed mechanical strength for process application, decreasing the cost and sometimes increasing the hydrogen flux.

Porous metallic support made, for instance, of sintered stainless steel (PSS), may represent a valid alternative to minimize the stress generated by the difference of thermal expansion coefficient between the membrane and the support, but the use of metallic support may cause significant inter-diffusion problems, which can be solved only by the application of an inter-diffusion barrier.

Actually, several types of barrier materials have been tested: TiN, TiO_2, Al_2O_3, α-Fe_2O_3, and YSZ; TiN was considered the most promising one, and sputtering seems to be the best application technique.

The composite Pd on PSS membranes prepared with an oxide barrier has been shown to be stable up to a temperature of 450°C for over 6,000 h [1].

Hydrogen permeance of 100 Nm^3/m^2 h $bar^{0.5}$ at 500°C can be achievable today; during tests on the industrial pilot plant (refer to Chap. 10), values up to 24 Nm^3/m^2 h $bar^{0.5}$ have been measured at 436°C; this value should double by operating at 500°C.

As far as the method of formation of thin Pd layers, electroless plating has given excellent results without showing any problem of hydrogen embrittlement (see ECN module experimental data); PVD sputtering represents a valid alternative for Pd alloy film, in order to overcome the difficulty to control the film alloy composition.

Furthermore, in general, the idea to manufacture the membrane components (film and the support) separately and then assembly together will allow to have a better quality control of the products, reducing the fabrication costs and making easier the assembling.

11.3.2 Interaction of Syn-gas Components with Pd and Pd Alloy

Absorption of CO, CO_2, and CH_4 on Pd surface has been reported in literature as a potential problem in reducing permeability or even for the complete deterioration of the entire membrane. The idea beyond is that such components may be absorbed on active surface sites blocking the hydrogen permeation.

The results reported by Ma et al. [1] on the effect of CO and CO_2 on composite Pd membrane was negligible in a hydrogen-rich environment of steam reforming of natural gas at 500°C; also, the authors' experience [3] under real operating conditions at a maximum temperature of 450°C did not evidence changes in permeability with a Pd membrane treating a syn-gas from a steam reforming step.

Sulfur poisoning is not a real issue because organic sulfur in the feed is normally converted and removed through a hydro-treating and ZnO adsorption prior of the reforming step.

11.3.3 Long-term Stability

In order to be viable for commercial applications at industrial level, membranes have to be tested for long periods to verify long-term stability over thousands of hours. Iaquaniello et al. [4] presented a competitive architecture for producing hydrogen based on membrane life assumption of at least 24,000 h of continuous operation.

Alternatively, the development of accelerated procedures for durability testing may shorten the required testing phase, as was proven by Bredesen et al. [5].

Data reported by Matzakos [6] showed a composite Pd/PSS membrane stable for approximately 6,000 h at 500°C under actual steam reforming conditions. Experimental data [4] with the ECN membranes over a period of more than 500 h, and not less than 50 cycles of heating and cooling, have not shown any evidences of instability problems. The rate of temperature changes used during the tests was of the order of 3–4°C per minute, in order to bring the module in operation within 2–3 h.

Although operating data have not yet confirmed a minimum life of 3 years under industrial operating conditions, it becomes quite evident that such a target is not so far away to be achieved. As far as the hydrogen purity is concerned, data collected at the pilot unit and presented in Chap. 10 show a hydrogen purity greater than 99.95% over 500 h of operation, with CO content in the range of 20–50 ppm.

As we said before, potential of membrane reactors is enormous, and in particular, they could play a key role with endothermic reactions because the higher reactant conversion at lower temperature allows less energetic and more efficient processes to be developed. As an example, natural gas steam-reforming is a rather expensive process, energy and capital expensive due to the endothermicity of the reaction.

A lower reaction temperature will make possible, for instance, to apply the concept of topping cycle cogeneration, where power is generated introducing a heat engine between the flames and the process fluid. Iaquaniello et al. [7] reported a specific application for the integration of a gas turbine with a fired heater.

A new configuration will then emerge with membrane reactors where heat and power are cogenerated, substantial energy savings and CO_2 emissions are achieved, and manufacturing costs are reduced.

Under a more general frame, the elimination of fuel combustion in fired heaters allows the use of multiple sources of low temperature heat as solar-heated molten salts or helium from a nuclear power [8] and not only the gas turbine exhausts. The cogenerative scheme presented for steam reforming could easily be extended to the other chemical processes that have been analyzed in this book.

11.4 Economical Perspectives

The major cost for a composite Pd-based membrane is not the Pd material being now feasible to produce films in the range of few μm. With such a thinner film, cost of Pd becomes less of factor. The cost of 3-μm thickness Pd layer plus cost for ceramic support tube was indicated at 900 €/m^2 by van Delft et al. [9]. Cost of the entire manufacturing process is the real determinant of the overall fabrication cost.

A reasonable price within 1,500–2,000 €/m^2 of thin Pd alloy membranes could be reached at the end of next decade only if a high cumulative volume of production is reached. Of course, such a cost has to be associated with a hydrogen permeance of at least 100 Nm^3/m^2 h $bar^{0.5}$ at 500°C.

A couple of preconditions also need to be fulfilled: first of all, long-term stability/durability has to be demonstrated without any doubt; secondly, one or two defect-free technologies need to emerge to sustain cost reduction based on economics of scale. In Chap. 3, Physical Vapor Deposition technology was indicated together with a roll-in process for producing the Pd-based thin film, a very promising technology for industrial mass production. Separation of Pd-based thin film preparation from the manufacturing of the support and from the final assembling may also emerge as the winning Membranes Manufacturing Strategy (MMS) in order to bring such a technology to the market.

11.5 Conclusions

The major features and application opportunities for integrated membrane reactor and staged membrane reactor configuration have been described in some details. We may conclude that membrane reactors actually show promises for conversion of equilibrium-limited reactions, selectivity toward some intermediates and improvement of energy efficiency. Consistent application of such technology requires long-term stability of the membrane, defect-free fabrication, a hydrogen permeance of at least 100 Nm^3/m^2 h $bar^{0.5}$ and a MMS able to gain from economy of scale when high cumulative volume of production can be achieved. It is also important to promote tests and results outside lab scale. What presented in Chap. 10 is only a first step toward industrial scale analysis and much more is required in terms of data reproducibility and technology acceptability. Although the primary role in developing such membranes and membrane modules belongs to material engineers, chemical engineers will play an important role in developing a better knowledge in modeling such complex unconventional reactors, on more heat- and energy-efficient schemes and on looking at new applications.

When such knowledge will achieve commercialization? Opinions are quite different in spite of the large number of patents and papers recently published on such a subject as we have seen in Chap. 2.

It is quite clear that any new invention takes time, in particular, in a very mature business such as the oil, gas, and petrochemicals. However, we are pretty confident that in the next 10–15 years such a technology will definitively emerge.

References

1. Ma YH, Mardilovich PP, She Y (1998) Proceedings, ICIM6, 246
2. Hydrogen from Coal R&D Plan, Office of Fossil Energy, June 10, 2004
3. Iaquaniello G et al Characterization of membrane modules for hydrogen separation in an industrial environment (submitted for publication)
4. Iaquaniello G, Giacobbe F, Morico B, Cosenza S, Farace A (2008) Membrane reforming in converting natural gas to hydrogen: production costs, part II. Int J Hydrogen Energy 33: 6595–6601
5. Bredesen R et al (2009) Cachet WP3
6. Matzakos A (2006) Presentation at the 2006 NHA, Annual Hydrogen Conference, 15 March 2006, Long Beach, CA
7. Iaquaniello G et al (1984) Integrate gas turbine cogeneration with fired heaters. Hydrocarbon Process 63:57–60
8. De Falco M, Barba D, Cosenza S, Iaquaniello G, Marrelli L (2008) Reformer and membrane modules plant powered by a nuclear reactor or a solar heated mol-ten salts: assessment of the design variables and production cost evaluation. Int J Hydrogen Energy 33:5326–5334
9. van Delft YC et al (2005) Hydrogen membrane reactor for industrial hydrogen production and power generation. 75th International Conference on Catalysis in Membrane Reactors, Cetraro, Italy, 11–14 September 2005

About the Editors

Marcello De Falco

Marcello De Falco got a master degree in Chemical Engineering at University of Rome “La Sapienza” in 2004; he got Ph.D. on “Industrial Chemical Processes” in April 2008, with the thesis “Pd-based membrane reactor: a new technology for the improvement of methane steam reforming process”. He is researcher for Engineering Faculty of Campus Biomedico of Rome from May 2010. His research activity is mainly focused on chemical reactors mathematical modeling, hydrogen production processes, solar technologies, cogeneration plants design. He taught “Chemical kinetics and chemical reactors” and “Chemical reactors” at the Faculty of Chemical Engineering of University of Rome “La Sapienza”. Presently, he teach “Transport Phenomena” at Engineering Faculty of Campus Biomedico of Rome. He is the author of 34 scientific papers (20 on international journals and 14 on congress proceedings) and of 3 chapters on international books about selective membrane technology and application.

Gaetano Iaquaniello

Gaetano Iaquaniello is a Vice-President of Technology and Business Development at Tecnimont KT SpA (process and engineering company), Rome, Italy. He received his M.Sc. in C.E. cum laude from the University of Rome in 1975, docteur d’université, the UER de Sciences, University of Limoges, France, in 1984 and an M.Sc. degree in Management from the London Business School, University of London, in 1997. He has more than 35 years of experience in designing and operating the chemical plants, particularly, for syngas manufacturing. After being a technical director, he is now a Head of the R&D activities and co-ordinates several national and European projects. He authored and co-authored numerous papers and patents on syngas production and more recently on membrane reactors.

Luigi Marrelli

Luigi Marrelli is full professor of "Chemical Reactor" and Dean of the Faculty of Engineering at University Campus Bio-Medico of Rome. His previous academic career was held at University of Rome "La Sapienza" where he was professor of Physical Chemistry and of Chemical Reactors and President of the didactic committee of Chemical Engineering. He is member of the International Research Centre on Sustainable Development (CIRPS) and of the Research Centre Hydro-Eco: Hydrogen as an alternative and ecological energy carrier (Sapienza University of Rome). For many years he worked in the field of university cooperation with developing countries realizing projects in Perù, China, Kazakhstan, etc. His scientific activity, summarized in over 100 papers and congress communications, pertains the fields of chemical thermodynamics, chemical and biochemical kinetics and reactor modelling. Over the last years he was involved in research projects on hydrogen production by steam reforming of methane in membrane reactors and from water by thermo-chemical cycles.

About the Contributors

Angelo Basile

Dr. Angelo Basile, Honours Degree in Chemical Engineering and Ph.D. in Technical Physics, is responsible at the Institute on Membrane Technology of the Italian National Research Council for the research related to the ultra-pure hydrogen production and CO_2 capture using Pd-based Membrane Reactors. He published more than 100 papers as first name and/or corresponding author; 26 invited speaker or plenary lectures and 190 papers in proceedings in International Conferences; lecturer in various Summer Schools on Membrane Reactors organized by the European Membrane Society; author of various international reports; editor of 5 scientific books and also other 27 chapters on international books on membrane science; 8 patents (of which: 1 at European level and 1 at worldwide level). He is referee for 43 international scientific journals and Member of the Editorial Board of 13 of them; member of 3 Int. Associations (International Association for Hydrogen Energy, European Membrane Society and American Chemical Society); guest editor of 6 Special Issues on Membrane Reactors. Since 2001, Basile is inserted as referee in the list of the Italian National Research Council named "Grande Albo dei Referee" and also was inserted in the list of the European experts in the field "Nanotechnologies and Nanosciences" of the VI Research Framework Programme (Brussels). Moreover, he participated to and was responsible of various national and int. projects on membrane reactors. Basile was the Director of the ITM-CNR during the period Dec. 2008–May 2009.

Adelina Borruto

Adelina Borruto is a Professor of Biomaterials and Metallurgy at the Engineering Faculty in the Department of Chemical Engineering and Materials, University of Rome "La Sapienza". She is a member of "Societe Francaise Metallurgy" and of "International Association for Hydrogen Energy". She participated in the Strategic Program: (1) FISR–MIUR "Pure hydrogen from natural gas by total conversion reforming, by integration chemical reaction and membrane separation", objective: membranes materials. (2) FISR–MIUR–TEPSI "Hydrogen production by thermochemical processes powered by solar energy: Sulphur-iodine cycle", objective: materials corrosion resistance in environments with hydrogen, iodine and hydriodic acid. She produced about 80 publications in national and international Journals. She is author of the book "Meccanica della Frattura", Hoepli 2002. She has an European patent in pending in USA: "Hip prosthesis and designing method thereof".

Elvirosa Brancaccio

Elvirosa Brancaccio is on the staff at Processi Innovativi S.r.l., L'Aquila, Italy a company owned by Tecnimont KT. She graduated in Chemical Engineering from the University of Naples in 2005 and received a MS degree in Energetic Engineering from the University del Sannio in 2006. She was awarded a Ph.D. in Chemical Engineering from the University of Salerno in 2010. She has worked for four years in the area of research and development of innovative products in the field of energy saving and her research is now focused on the study of hydrogen production from hydrogen sulphide by catalytic reaction. She has co-authored a number of articles, research reports, international conference presentations and one patent.

Paolo Ciambelli

Full professor in Chemical engineering at University of Salerno from 1990. Director of NANO_MATES (Research Centre for Nanomaterials and Nanotechnology at University of Salerno) from 2008. Chairman of the Board for Doctorate in Chemical Engineering at University of Salerno from 2006. Member of Senato Accademico at University of Salerno from 2003 to 2009. Head of Department of Chemical and Food Engineering from 1996 up to 2002. Member of the Research Council at the University of Salerno from 1996 up to 2002. Member of American Chemical Society (ACS) and of American Institution of Chemical Engineering (AICHE) from 1997. Member of Scientific Council of CUGRI (Inter University Consortium of Risk and Hazard management) from 1999 to 2005. Member of Director Board of Catalysis group of Italian Chemical Society from 2000 to 2002. Member of Director Board of Division of Industrial Chemistry of Italian Chemical Society from 1998 to 2000 and from 2004 to 2009. President of Italian Association of Zeolites (AIZ) from 2000 up to 2003. The main research activities of Prof. Ciambelli are in the field of Industrial Catalysis: catalytic processes for the abatement of pollutants from fixed (power plants) and mobile sources (lean-burn gasoline, diesel, natural gas fuelled engines) and for hydrogen production (autothermal reforming). Further subjects are in the field of photocatalysis, properties and applications of nanomaterials (carbon nanotubes, nanoparticles, zeolites), leather tanning industry, catalytic combustion, coal blends combustion, process safety. Prof. Ciambelli was responsible of National and European Research projects and of various Research Contracts financed by public institutions and private companies. He is author of over 250 publications on international journals and 8 patents.

Jan Galuszka

Jan Galuszka, is a Senior Research Scientist at Natural Resources Canada, CanmetENERGY in Ottawa, Canada. He received his M.Sc. in Chemistry (with distinction) in 1970 and his Ph.D. also in Chemistry, in 1973, both from the Jagiellonian University, Cracow, Poland. He has close to 40 years of experience in various aspects of catalysis, surface spectroscopy, natural gas conversion, Fischer-Tropsch synthesis and membrane development and application. He managed many R&D projects and authored and co-authored numerous publications, reports, patents and international conference presentations.

Terry Giddings

Terry Giddings is a Research Chemist at Natural Resources Canada, Canmet-ENERGY in Ottawa, Canada. He received his B.Sc. in Chemistry in 1985 at Carleton University, Ottawa, Canada. He has over 20 years of experience in the area of membrane research and development. Most recent work includes the development of hydrogen selective silica membranes as applied to the partial oxidation of methane, decomposition of H_2S, dehydrogenation of alkanes and the water–gas shift reaction. He co-authored a number of publications and research reports.

Adolfo Iulianelli

Adolfo Iulianelli, Degree in Chemical Engineering and Ph.D. in Chemical and Material Engineering, is expertise at the Institute on Membrane Technology of the Italian National Research Council for the research related to the ultra-pure hydrogen production through the technology of Pd-based Membrane Reactors as well as to preparation and characterization of polymeric membranes for PEM fuel cell applications. He published 30 papers in International ISI Scientific Journals, 1 patent and more than 30 publications in proceedings for International Conferences as poster or oral presentations. He was different times invited speaker in International, National Conferences and Training Schools on Membrane Reactors. Author of various chapters of books on Membrane Reactor Technology and PEM fuel cells. Editor of a scientific book for International Journal of Hydrogen Energy and even reviewer for more than 10 International Scientific ISI Journals. Moreover, he participated to various national and international projects on membrane reactors and PEM fuel cells.

Dina Katzir

Dina Katzir is Deputy CEO and Head of R&D at Acktar Ltd, Israel. She holds M.Sc. Degree in Applied Physics and has close to 20 years experience in various aspects of material science and optics. She has authored of above 50 patents and numerous papers. She has published on a wide range of subjects: IR optics, thin film coatings, membranes, medical devices, electronics and solar energy.

Simona Liguori

Simona Liguori received degree in Chemical Engineering at the University of Calabria in 2008 with a thesis on "Experimental analysis on ethanol steam reforming. Comparison between traditional and membrane reactors". Currently, she is Ph.D. student and the topic of her study is "Experimental & simulative analysis on the steam reforming of biofuels in both membrane and fixed bed reactors". Her scientific activity focuses on the use of renewable sources such as biofuels produced from biomass instead of derived fossil fuels for producing hydrogen via reforming processes in alternative device as the membrane reactor. She published 7 papers on International Refereed Journals, more than 8 papers in Proceedings in International Conferences and she is referee for International Journal of Hydrogen Energy.

Erika Lollobattista

Erika Lollobattista is a Chemist, actually working in the R&D group of Processi Innovativi Srl, a company owned by Tecnimont KT SpA, Rome, Italy. She received his Master degree in Chemistry in 2003, from the University of Rome "La Sapienza". Most recent work includes the study and development of hydrogen selective membranes applied in methane steam reforming and water–gas shift reactors. She managed/attended a number of European research projects in the field of membranes application.

Tiziana Longo

Tiziana Longo, Degree in Chemical Engineering and Ph.D. in Chemical Engineering, is researcher at Institute on Membrane Technology of the National Research Council (ITM-CNR), located at the University of Calabria, Rende (CS). She is involved in the research field related to the ultra-pure hydrogen production by reforming reactions of biofuels carried out in membrane reactors as well as the characterization of inorganic membranes to gas and vapor permeation. She published 11 papers on International Refereed Journals and more than 20 papers in Proceedings in International Conferences. She was a lecturer in the Second Training School on "Sustainable Hydrogen Production and Energy" organized by the COST Action 543 and she is referee for International Journal of Hydrogen Energy.

Giovanni Narducci

Giovanni Narducci is a Mechanical Engineer. He actually cooperates with the University of Rome "La Sapienza"-Department of Chemical Engineering Material Environment. He conferred a Master degree with honours in Mechanical Engineering at the University of Rome "La Sapienza". His recent works include the study of hydrogen selective membranes and their characterization by Scanning Electron Microscopy (SEM) and Energy Dispersive Spectroscopy (EDS).

Vincenzo Palma

Vincenzo Palma, associate professor of Industrial Chemistry at Department of Industrial Engineering of University of Salerno (Italy). He was graduated in Industrial Chemistry at the Faculty of Mathematics, Physics and Natural Science of the University of Naples in the 1991 with maximum grade. From the 1995 to the 2005 he was assistant professor of Industrial Chemistry and from 2005 he is associate professor at the university of Salerno. In the last 20 years he studied in the field of heterogeneous catalysis and accumulating experience in the field of catalytic environmental pollutants abatement and syngas production. He is scientific responsible for R&D projects and authored and co-authored numerous publications, reports, patents and international conference presentations.

Emma Palo

Emma Palo is a chemical engineer actually working for Tecnimont KT S.p.A., Rome (Italy) in R&D. She graduated in Chemical Engineering cum laude at the University of Salerno in 2003. She received her Ph.D. in Chemical Engineering from the University of Salerno in 2007. From 2008 to 2010 she was a research fellow at the Dept. of Chemical and Food Engineering of the University of Salerno, with a research activity mainly focused on the study and application of heterogeneous catalysis in energy and environmental fields, including the syngas production from hydrocarbons and renewable sources, syngas purification, and abatement of nitrogen oxides through selective catalytic reduction with methane, for applications in the cars engines field. She is co-author of a number of publications, research reports, international conference presentations and one patent.

Annarita Salladini

Annarita Salladini is a chemical engineer actually working for Processi Innovativi Srl, a process company owned by Tecnimont KT SpA, Rome, Italy. She received her M.Sc. in C.E. cum laude in 2004 and her Ph.D. in 2009, both from the University of L'Aquila. She has been a teaching assistant in the area of Chemical Plants Process and Design at University of L'Aquila. She was involved in research project regarding renewable and energy saving technologies and co-authored publications and conference presentations. Since 2009 she has been involved in construction, commissioning, start up and operation of a reformer and membrane modules pilot plant.

Moshe Sheintuch

Prof. Sheintuch is a Beatrice Sensibar Chair Professor in Environmental Engineering with an appointment in the Faculty of Chemical engineering and a secondary appointment in the Faculty of Civil and Environmental Engineering. He is in the Technion since 1977, after graduating with Ph.D. from the University of Illinois. He spent three sabbaticals as a visiting faculty in University of Houston and Princeton University (twice). His Chemical Reaction Engineering and Environmental Catalysis group have been working on catalytic, biochemical and other reactors and has published more than 200 papers in international journals in Chemical Engineering, Physics and Physical Chemistry. His experience in catalysis include oxidation and hydrogenation reactions in the gas and liquid phases. He has published extensively on the subject including several experimental papers on dehydrogenation reactions in membrane reactors. Other areas of his expertise include reactor dynamics, catalytic dynamics, mathematical modeling and DFT calculations. He is currently the Deputy Senior Vice President of the Technion.

David S. A. Simakov

David S. A. Simakov has graduated in 2004 at the Technion—Israel Institute of Technology with M.Sc. degree in Chemical Engineering and he has received his Ph.D. at the same faculty in 2010. The topic of his M.Sc. thesis was fabrication and characterization of ceramic nanomaterials. He has completed his Ph.D. project in the area of chemical reaction engineering, focusing on development of the catalytic membrane reactor for hydrogen production. He has joined a start-up company developing a new generation of polymer membrane fuel cells as a R&D Chemical Engineer during 2008–2010. Dr. Simakov has published several scientific papers on the subject of fabrication of ceramic nanopowders, membrane reactor dynamics and optimization of hydrogen production in catalytic membrane reactors. His experience in membrane reactors includes mathematical modeling, design and experimental investigation of hydrogen generation systems and their optimization. Other areas of his expertise are heterogeneous catalysis, reactor dynamics, fuel cells and computational system biology.

Index